MAINE
BIRDS

Selected Articles from
MAINE FISH AND WILDLIFE
Magazine

Compiled by

NORTH COUNTRY
PRESS

UNITY, MAINE

Material © Maine Department of Inland Fisheries and Wildlife.
Reproduced by permission.

Library of Congress Cataloging in Publication Data

Maine Birds.
    1. Birds—Maine—Addresses, essays, lectures.
I. Maine fish and wildlife. II. Thorndike Press.
QL684.M2M34   598.2'9741   78-9702
ISBN 0-945980-11-6 (formerly ISBN 0-89621-010-3)
Third printing.
First North Country Press printing, 1989.

INTRODUCTION

The articles in this publication have been arranged into three categories—upland game birds, migratory game birds, and non-game birds.

The distinction between "game" and "non-game" species is made here because the publication from which these articles are reprinted—MAINE FISH AND WILDLIFE Magazine—is directed primarily toward sportsmen, who find the distinction important to their understanding of the interaction of the various species from a management standpoint.

Although there will be found in these articles some mention of wildlife management techniques, the reader will find that the primary focus of the text is on life histories of the species. Many articles of a more scientific nature have appeared in MAINE FISH AND WILDLIFE Magazine during the nearly 20 years of publication, and these can be obtained from the Maine Department of Inland Fisheries and Wildlife, 284 State Street, Augusta, Maine 04333.

The Editor

Freeport Community Library
Library Drive
Freeport, ME 04032

DATE DUE

APR			
FEB 0 4 2003			
FEB 1 8 2003			
GAYLORD			PRINTED IN U.S.A.

TABLE OF CONTENTS

PART I: UPLAND GAME BIRDS

Maine has only two species of bird which could be called native upland game birds—the ruffed grouse and the spruce grouse. The ring-necked pheasant has a stable population in Maine, but they are here today because of intentional introduction some years ago, and because of an ongoing raising and stocking program.

Some confusion may arise, also, because of the inclusion of woodcock, rails, and snipe into the first article of this section. This is easily explained. This article, Upland Game Birds, is written more from a standpoint of habitat type, while the distinction between upland game birds and migratory birds is more commonly made in reference to whether or not the bird actually does migrate for the winter months.

For this reason, articles on woodcock and snipe will be found in the second section of this book, Migratory Birds, although they are mentioned at the start of this section.

© Leonard Lee Rue III

TABLE OF CONTENTS

Ring-necked pheasant

Upland Game Birds

By Richard B. Parks

O N AN AUTUMN DAY when the hardwoods are at their peak in color, ranging from a bright yellow through orange to a deep scarlet, nothing can give the bird hunter more pleasure than going afield with his pointer or retriever in quest of one or more of the species of birds classified as "game birds" by our laws. In general, these are the species that are hunted for food or sport and for which there is an open season.

Not all hunters who roam the fields, forests, and marshes of Maine are familiar with what the biologists call the "life histories" of these birds. This term means the nesting habits, food habits, cover requirements, and range of the individual species. The hunter, afield with his favorite dog and shotgun, knows the likely fall cover in which to find his sport, and if he knows its habits in other seasons of the year, perhaps he can better understand his quarry and the laws governing its hunting.

Game birds fall into two broad classifications — resident, which are essentially birds that "stay put" all year around, and migratory birds, which either spend only a season or two here in Maine or are merely passing through our state on their way to their wintering grounds or breeding grounds.

The first three birds which we shall briefly discuss represent the resident game birds. The state has complete jurisdiction over the laws pertaining to these species. The remaining birds fall into the migratory category and are governed by federal regulations.

T O MANY, the king of all game birds in Maine is the ruffed grouse or "partridge" as it is commonly called by most Maine hunters. This is a woodland bird found throughout Maine at all seasons of the year.

Grouse usually nest in wooded uplands, but occasionally in low, damp woods. The nest is ordinarily a depression in the soil, lined chiefly with dead, forest leaves, and hidden by an overhanging limb or concealed by a log. Grouse have been known to nest in trees in the abandoned nest of a crow, but this is a very rare instance.

Usually there are from seven to fourteen eggs in the nest, and the young birds hatch after twenty-four days incubation. Nesting usually occurs in late April, May, and June, and each female has a single brood yearly. One may readily see how vulnerable the grouse nest is, not only to predators which roam the ground but also to very cold or wet weather during the nesting season.

The food of the ruffed grouse in the fall is varied but usually consists of mast like hazel nuts, beech nuts, and acorns, as well as fruit such as the chokecherry, apple, checkerberry, grape, and thornplum. Leaves of many herbaceous plants are also taken. In winter, buds of trees including birch, willow, apple, beech, and maple are freely taken. Spring and summer food consists largely of insects and young growing plants like grass and clover.

When this country was first settled, the grouse was known as the "fool hen"; easily killed with sticks and stones, it was considered fair game for small boys. The survivors have become "educated" over most of their range and resort to tricks such as swinging behind a tree when flushed, running quietly until beyond gunshot range, and leaving the woods and alighting in a field while the hunter and his dog search the thickets in vain.

T HE CANADA GROUSE or Canada spruce partridge has all but disappeared from the inhabited sections of Maine. It seems to be holding its own fairly well in the more remote regions of the state although it is not

The ruffed grouse or partridge is the king of game birds in Maine. His explosive takeoff helps make him an elusive target for the best of hunters.

common even there. In 1927, this bird was listed as uncommon-to-rare in Somerset, Franklin, Oxford, Knox, Piscataquis, Penobscot, Waldo, Hancock, and Washington counties, very rare in southern Maine, but less uncommon in Aroostook County. This species is also a ground nester, preferring to nest under brush at the base of a tree in swampy, coniferous woods. The nest is of twigs, leaves, and grass and contains eight to sixteen eggs, usually twelve. Incubation takes about seventeen days, and the young are hatched in late May or early June. There is one brood per year.

Food of the spruce grouse consists largely of foliage and buds of spruce, larch, and fir in winter; in spring and summer, the bird readily eats insects, including grasshoppers in large quantities, as well as the tender parts of low growing herbaceous plants. In the fall the principal food is wild berries. The young are said to subsist mainly on insects and spiders.

These birds are usually quite tame and an easy mark for the hunter because they live in a greatly restricted range well away from most human habitation and have not become "educated" as have the ruffed grouse. Some escape the hunter by retiring to dense, coniferous thickets and refusing to fly unless almost stepped on. The flesh of these birds is usually impregnated with the taste of spruce, making the table quality rather poor.

THE RING-NECKED PHEASANT, believed to be a native of Asia, has been introduced into Europe and the United States. Several other species of pheasant have been bred, either in captivity or as a semi-domestic fowl, but only one has become acclimated to the wild state in the United States. Early records of pheasants in Maine date back to 1897, but it is possible some birds were within our borders before then from stockings in Massachusetts and New Hampshire. Small numbers were stocked in Maine, primarily in the southern counties, between 1897 and 1933, but no open season was allowed until 1935. In 1933, the first pheasants were liberated from the state-owned game farm at Dry Mills. This farm was authorized by the 1931 Legislature.

The ring-neck is a bird of open, agricultural lands, especially where grain farming is heavily practiced. In Maine, they do best along the coastal belt where truck gardens, salt marshes, and farmlands are found. Woodland is not good pheasant habitat.

Most of Maine's pheasants are raised artificially by incubators, from eggs obtained from breeding stock kept at the game farm. In the wild, the birds usually nest in brushy pastures or fields of grass, rarely in the woods. Usually the nest is on the ground and is made of leaves, grass, and straw. The usual six to twelve eggs are incubated twenty-three to twenty-five days, and the clutch is hatched some time between late April and late June.

Pheasants are primarily seed eaters and ground feeders. Insects, especially grasshoppers, are frequently taken during the summer, as are snails, slugs, and earthworms. Other summer foods include various tender green leaves, fruit, and grain. Grain and seeds form the principal fall foods when the berries and fruits are gone. Winter food is grain and weed seeds when available; some buds and garbage are taken as last resort.

As noted, the pheasant is a ground feeder and, as such, is limited in Maine because of heavy snow cover, making food difficult or impossible to obtain. Some pheasants do "winter over" in Maine, especially in areas where they are artificially fed or where they can obtain food themselves. The latter areas include farms where fresh manure (in which weed seeds can be obtained) is put outdoors and tidal areas along the coast where snails and other animal matter can be obtained, as well as seeds.

THE AMERICAN WOODCOCK, or timberdoodle, as he is sometimes called, is a most important migratory upland bird in Maine. Some sportsmen elevate this small bird above the ruffed grouse as a favorite game bird. Although the woodcock is hunted to some extent throughout the state, habitat conditions make eastern Maine, including Washington and Hancock counties, a superior area in which to hunt them.

The woodcock is found both nesting and migrating in all sections of Maine. However, they are not abundant in the heavily forested parts of western and northern Maine. They prefer reverting farmland and burns which are growing up to alders and gray birch with some open land between these alder runs and gray birch patches.

This bird's breeding range covers almost all of the eastern United States, with the exception of a narrow belt along the Gulf Coast and in southern Florida. Its breeding range extends just west of the Mississippi River, so it is truly an eastern bird. The breeding range also extends into southern Ontario, southern Quebec (but only about as far north as the St. Lawrence River), New Brunswick, Nova Scotia, and extreme southern Newfoundland. The wintering range is much more restricted, being a belt about 400 miles wide from Virginia to Texas, along the Atlantic and Gulf coasts. The great majority of birds, however, winter in Louisiana and western Mississippi, a very restricted part of their winter range.

The woodcock nests in a shallow depression on the ground and does not build much of a nest. Often there is very little attempt made at concealment as the bird itself blends well with its surroundings. Usually four eggs are laid in late April or very early May, and incubation requires twenty-one days. The young can take short flights fifteen or sixteen days after hatching and are fully grown at five months of age.

Earthworms constitute more than four-fifths of this bird's diet. The bird is well equipped for probing moist earth for worms by having a long, flexible bill. More than 95 per cent of the woodcock's diet is animal matter — earthworms, beetle and other insect larvae, and spiders. Vegetable matter in the diet includes raspberries, blackberries, and the seeds of sedge and alder.

The woodcock is one of the species of which the female is larger than the male. Both are roughly robin-size. The male generally weighs about six ounces and the female averages seven or eight ounces, and both have a bill about 2½ inches long. The birds are crepuscular; that is, they are most active at dawn and dusk — one reason why they are seldom seen by the casual observer. A good bird dog is a great asset in looking for woodcock.

THE WILSON'S SNIPE or jack snipe is another migratory game bird of Maine although perhaps not as well known or as sought after as the woodcock, but nevertheless a fine game bird. Its zigzag flight makes it a difficult target for even the most proficient shooter.

The snipe weighs about four ounces and has a bill about 2½ inches long. This bird is found nesting and migrating throughout Maine. Its range includes all of North America, north to the limit of trees, as well as parts of Central and South America. During migration, snipe are more common along the coastal belt than inland. They prefer open bogs and wet meadows and are often found in salt meadows during migration.

Snipe nest in shallow depressions in grass or sedge near or in a bog or wet meadow, and make very little effort at concealment. Usually four eggs are laid in May, and incubation takes about twenty days.

Like its close relative, the woodcock, most of the snipe's diet consists of animal matter. Earthworms, crustaceans, and insects, including grasshoppers, click beetles, cutworms, wireworms, mosquitoes, and water beetles, make up the bulk of its diet. Seeds of smartweed and other vegetable matter are also taken in small quantities.

For a number of years, this bird was fully protected by federal law because of greatly diminished numbers. For the last decade or so, regulations have been relaxed as the snipe had increased sufficiently under full protection.

RAILS, although not commonly hunted in Maine, are another of our migratory game birds. Several species are found at times in Maine but only two are considered common, the sora rail and Virginia rail.

The range of these birds is generally over all of North America, from the northern limit of trees, south into Central America and northern South America. The rail generally nests in the northern part of this range. The wintering grounds extend from northern California through Illinois and to South Carolina. The Virginia rail occasionally nests in Maine, being most common in eastern Maine. The sora rail commonly nests in Maine and is particularly fond of cattail swamps for its nesting site.

These birds are marsh inhabitants and are about the size of a starling. The Virginia rail may be easily identified by its long, slightly decurved bill. The sora rail has a short, yellow, chicken-like bill. Food consists of seeds of aquatic plants (wild rice is a favorite), insects, and snails.

The rail nests in fresh water marshes or along brushy banks of marshy streams. The sora, however, sometimes uses grass or grain fields. Both species make their nests in weeds, grasses, dead stalks, or a pile of broken down reeds, generally on a hummock in the marsh, and generally well hidden. Usually four to a dozen or more eggs are laid in mid May to early July. Incubation takes about fourteen days.

Rails are weak fliers and are about as difficult to shoot as a tin can floating down a rapid current. In the mid-Atlantic states and the South, they are hunted rather extensively. Usually one man will pole a boat through beds of aquatic vegetation, with a man in the bow doing the shooting. Although slow fliers, some birds turn and swing about the boat, making shooting more difficult and quite dangerous for the boatman. ∎

Among the best camouflaged of birds is the American woodcock, increasing in popularity with Maine hunters.

Notes on the history and
living requirements
of a popular game bird

The Ring-necked
PHEASANT

By Richard B. Parks

A S HE HAS with many good things to eat, man has known the pheasant a long time. Remains of ancient campfires show that primitive man was familiar with the pheasant before the time of recorded history. The latter tells us that pheasants were brought to Europe from the Caucasus in the tenth century B.C. In the watershed of the Phasis River (now Rion River), the birds became abundant and from this geographical area came the name "phasian bird" or "pheasant." Science used the Latin term *Phasianus* for the genus; and from *Colchis*, the name of the province in which the Phasis River flows, the specific name of the common pheasant was derived. Thus, the technical name of *Phasianus colchicus* designates our common pheasant.

Through man's efforts over the centuries, many wild birds and animals have become domesticated, but nowhere has the pheasant become truly so. This very quality has earned it widespread transportation and propagation as a sporting bird. Hunting does not make the pheasant wild, but rather, its own natural characteristics make it so. It is eagerly sought by thousands of hunters and has spread to all suitable parts of the world.

Pheasants were introduced in the United States by Governor Wentworth on his estate in New Hampshire in 1790. Richard Bache, son-in-law of Benjamin Franklin, also released birds, at about this same time, on the shore of the Delaware River in New Jersey. Neither of these stockings was successful. The first known successful stocking was made in the Willamette Valley in Oregon in 1881. These birds multiplied rapidly, and a shooting season was declared in 1892. The reports show that fifty thousand birds were killed the first day.

Early records indicate that pheasants were released in Maine in 1897. However, some probably moved into southwestern Maine earlier, from liberations in New Hampshire and Massachusetts. These 1897 releases were at Searsport and at Dixon Island in Rangeley Lake. The island birds disappeared in the winter of 1899-1900. By 1912, pheasants were seen regularly around Portland and by 1918 had reached the Lewiston-Auburn area. The 1920 Inland Fisheries and Game Commissioner's report states that "the ring-necked pheasant has been reported this year in quite large numbers in York County and other southern portions of the state." By 1931, pheasants had been stocked in all coastal counties of Maine and some inland areas.

P HEASANT FOOD and pheasant cover are inseparable in the Northeast. Food should be in close association with protective cover to avoid undue exposure of the birds to enemies. Most northeast farms, including those in Maine, have enough cover for nesting, broods, roosting, and what we call loafing cover, which is merely heavy cover that protects the birds from predators.

Winter food, however, is generally insufficient. Such wild seeds and fruits as thornapple, apple, skunk cabbage, ragweed, burdock, grapes, and staghorn sumac provide some food over critical periods of the winter, but these natural foods should be supplemented by grain, especially corn. Other cultivated grains are readily eaten, but in this area they are grown in only small amounts or are not available during periods of deep snow.

Wild plants in Maine have all the qualities necessary for shelter or cover but are not always where needed or in the quantity needed. Swamp edges and depressions occupied by cattails and marsh grasses are preferred cover. Heavy growths of weeds are also frequently used. Important winter cover is pasture juniper and coniferous growth in its younger stages. All these types serve as escape cover in time of danger. Needless to say, the farther good cover is removed from food sources, the less valuable it becomes.

Breeding cover is usually in open country in brushy pastures, in fields of grass or grain and, rarely, in the woods adjacent to open land. Pheasants are ground nesters, building a nest of leaves, grass, straw, or similar material in which usually six to twelve olive brown to pale blue eggs are deposited. Nests have been found that contain as many as sixteen eggs, but this is a rarity. The female incubates her clutch twenty-three to twenty-five days.

The pheasant is polygamous, and a single cock is likely to have a harem of several hens. The cocks are vigorous and powerful, and, like the males of other polygamous species, they fight viciously for the possession of their harem.

Here in Maine, the year to year carryover has been determined at about 3.7 per cent of birds stocked, indicating that birds breeding in the wild form only a small portion of the shootable population found in the fall by Maine hunters. At the present time, upwards of thirty thousand pheasants are stocked annually by the Department of Inland Fisheries and Game. ∎

THE RUFFED GROUSE

Forest Drummer

By Henry S. Carson

THE RUFFED GROUSE, sometimes called partridge or birch partridge, is the most popular game bird in Maine. More people in Maine hunt this species than all other species of wildfowl combined.

The ruffed grouse is found in most of northern North America and is distributed through all of Maine. It is a bird of the deciduous and mixed deciduous-coniferous forest margins and second growth, or deciduous growth habitats associated with northern coniferous forests. It is especially suited to the climate and land uses in Maine. Lumbering has created ideal habitat for grouse, and abandoned farm lands and old burns are also of great importance.

Distinct rufous and gray color phases are recognized for ruffed grouse although various intermediate colors are also found in Maine. The bird's color above is generally brownish and spotted; below, it is more yellowish, barred with dark. Coloration is similar in both sexes. The fan-shaped tail and the wings are of equal length. Adult males weigh about 1¼ pounds and occasionally weigh up to two pounds. Females are smaller and weigh about one pound or slightly more.

Courtship activity in grouse is especially outstanding and takes place during early spring. The male grouse selects a log (or occasionally another object such as a stump, root, or rock) and begins his drumming. The beat of the bird's wings against air — not against the body or log — creates a "boom——boom-boom-br br br br-rr" drumming sound that is designed to attract the female grouse.

The male ruffed grouse can also puff his feathers out until he resembles a porcupine's back with all quills erected. The ruff, from which the bird gets its name, consists of two small patches of black feathers — one on each side of the neck. This ruff is especially noticeable on the male when he puffs his feathers. This demonstration is usually reserved for an audience of two or more females. Considerable strutting with tail fanned out is also done. Drumming activity is at a peak during May, is occasional during the summer months, and reaches a secondary peak during the fall. Drumming activity other than in the spring of the year has no relationship to mating.

GENERALLY IN MAY, the hen grouse makes a shallow ground nest out of old leaves, a few feathers, and debris. The nest may be at the base of a tree or in dense underbrush, usually near an old road or opening in the forest. She lays an average of eleven whitish to pale-brown eggs and incubates them for twenty-three or twenty-four days. Within a few hours after hatching, chicks leave the nest.

The hen grouse is a good, protective mother and will act belligerent or crippled at the approach of man, as well as hovering the chicks during heavy rains or hail storms. Cold wet weather is a health hazard to young grouse.

At this age the chicks feed on insects. By the time they are two weeks old, they can make short flights. By the last of September, when broods usually disperse, most of them have attained adult size.

Ruffed grouse eat many kinds of food. During the summer, insects are of great importance in the diet and

are supplemented by leaves and berries. Fall foods consist of clovers, berries, fruits, seeds, various leaves, nuts, and buds. During the winter months, buds of birch, aspen, and hazel are favorites. Spring is a transition period, and bud diets are augmented by available green leaves.

The ruffed grouse is a hardy bird and often burrows under the snow to escape bitter cold winter weather. Winter food is no problem, because the birds have the ability to eat buds from trees. Even after a sleet storm, ice-free buds, especially of beaked hazel, are available under protective softwood growth.

It is believed that no well defined grouse cycle exists in Maine, as periods of abundance and scarcity — real or seeming — do not occur simultaneously in all sections of the state. Hunting pressure succeeds only in making the grouse more wary and has little effect on the population.

Of the five leading grouse states, it is interesting to note, Maine has fewer hunters, fewer grouse harvested (total or per square mile), and the shortest hunting season. The other four top grouse hunting states are Michigan, Wisconsin, Minnesota, and New York. The grouse hunter in Maine does enjoy the highest hunting success, and he should be well satisfied with his lot. ■

The Spruce Grouse

DWELLER OF THE DARK FOREST

By Henry S. Carson

THE SPRUCE GROUSE or spruce partridge is not as well-known as its more popular cousin, the ruffed grouse. This is due to a much smaller population of birds living in more remote areas although suitable habitat for spruce grouse does exist in most of Maine with the exception of the southern end of the state. It is a bird of the spruce-fir forest and frequents shaded, mossy areas, often near a black spruce bog. In comparison with the ruffed grouse, it is more a wilderness bird. Ruffed grouse are believed to be at least 20 times more numerous than spruce grouse.

The sexes are colored differently, males being gray and black above; and black, black-and-white, and gray-and-white below. Females are gray and brown above, and brown and white below. Males have a patch of unfeathered red skin over each eye. At a distance, a female spruce grouse can be mistaken for a ruffed grouse, but examination of tail feathers is a satisfactory means of distinguishing between the two. Spruce grouse tail feathers have a reddish brown tip, but the tail has no wide, black band like that of the ruffed grouse. Spruce grouse are slightly smaller than ruffed grouse, weighing about one pound each.

Courtship activity takes place during early spring. The male "drums" by beating the air with his wings, while walking or flying. In comparison to a ruffed grouse, there are fewer drumming beats, and the sound is less audible. Vocal calls are also made by the male. Drumming territory is somewhat variable from day to day. The male can puff out his feathers for display purposes as the female can while protecting her brood.

The female spruce grouse builds a nest on the ground under low conifer branches, brush heaps, or tamarack. An average of 11 eggs are laid in a nest constructed with dry twigs and leaves and lined with moss and grass. The eggs are buff to reddish brown in color and are blotched and spotted with different shades of darker brown. Young hatch out during late June, and by the last of September most of them have attained adult size.

The diet of the spruce grouse is largely composed of coniferous needles—tamarack and fir in particular. Insects, berries, buds, leaves, seeds, moss, lichens and fungi are also eaten during the course of a year.

The male spruce grouse is identified by its grey, black, and white coloration and the red patch of unfeathered skin over each eye.

Spruce grouse are very tame, and it is usually possible to walk within 20 feet of one before it flushes or walks away. They have been killed with sticks and stones and even snared around the neck when roosting in a tree. Unlike the ruffed grouse, the spruce grouse never seems to gain any wildness by being hunted. The local name "fool hen" appears to be an appropriate description of this bird.

Much has been said about the quality of spruce grouse flesh, and opinions have ranged from "unfit for human consumption" to "very delicious." Actually, the meat *is* fit for human consumption, and the quality is a matter of taste and preparation. Young-of-the-year spruce grouse have only slightly darker flesh than ruffed grouse and the flavor is only a little stronger. The flesh of adult spruce grouse is dark purple in color, and it does have a strong, "gamey" taste. This wild flavor can be minimized by clean killing and immediate cleaning of the bird, or it can be accentuated by the opposite procedure.

The future for the spruce grouse in Maine is neither bright nor glum. Although the bird is very tame, it does live in dense coniferous areas that are usually avoided by ruffed grouse hunters. Unless land use practices change drastically in Maine, which is unlikely, there appears to be little danger of exterminating the dweller of the dark forest.

PART II: MIGRATORY BIRDS

The articles in this section concern those birds which migrate during the winter months, and for which Maine has a hunting season—the migratory game birds.

The reader will find articles concerning some other migratory birds in the third section of this book, Non-Game Birds, which deals with native Maine birds which are not hunted.

TABLE OF CONTENTS

EIDER DUCKS, ISLANDS, AND PEOPLE

A drake eider remains on guard while his mate selects a nest site.

Photo by Richard Ferren for Maine Cooperative Wildlife Research Unit

By Howard L. Mendall

WOULD YOU LIKE to own an island on the Maine coast? If so, you have a lot of company. The increasing demands on the real estate market for recreational land and the prices such property are bringing have been amazing to most Maine residents.

Objectives for owning islands include subdivision and development for vacation homes, industrialization, preservation for scenic or historic values, and establishing of picnic areas and camp grounds for the boating public. Also included are wildlife interests — the safeguarding of the breeding and resting sites of seabirds, waterfowl, and marine mammals that are as much a part of the coast as the surf itself.

Many of the birds that frequent the islands nest in large numbers in colonies—sometimes several hundreds or even thousands on a single island. Colonial nesting is a trait of gulls, terns, petrels, puffins, and razor-billed auks. Among Maine waterfowl, only the eider duck is a colonial species. These birds characteristically shun the mainland for breeding, in favor of islands which usually provide much safer nest sites.

The American eider is of particular interest to both nature lovers and waterfowl hunters of Maine. It is the largest duck in North America. The strikingly colored black and white drake averages around 4½ pounds, and the brown or buffy female is about a pound lighter. Eiders are gregarious at all seasons of the year, and flocks of several thousand during fall and winter are commonly seen at many places along the Maine coast.

Well adapted to rigorous weather and low temperature, eiders seldom come ashore except during the breeding season. They are able to ride out even a northeast blizzard in a cove on the lee side of a large island. They obtain nearly all of their food by diving and are skilled at underwater maneuvers. Numerous invertebrates make up their diet, with blue mussels, periwinkles, and sea urchins especially favored.

EIDERS ARE AMONG the earliest of the seabirds to nest. By the last week of April, in most years, pairs go ashore, the drake following his female while she selects the nest site. Nests may be in a variety of places: unconcealed in an old gull nest, in grass or weeds, or within a patch of raspberry or bayberry bushes. Sometimes the birds select nest sites under driftwood or boards or even a fisherman's abandoned shack. On wooded islands, eiders may pick a site under a tangled blowdown or beneath the drooping branches of a spruce. However, use of forested islands in Maine is the exception rather than the rule. Apparently, the preference is for small islands having only grass, herbs, or shrubs as cover.

Since eiders are colonial, they nest with others of their kind, sometimes as many as 200 or more pairs per acre. Moreover, they usually choose islands that have large populations of gulls. This seems strange, as both the herring gull and the black-backed gull are well known to be egg eaters; and, in addition, the latter is especially fond of ducklings. Yet, we commonly find an eider nesting only three feet from an incubating gull. The fact that eiders appear attracted to gulls may have more value to the eiders than some observers believe. Gull colonies always contain non-breeding or immature birds. With no nests of their own to defend, these gulls are on the prowl for easy food. Such birds, together with crows and ravens (which are even more addicted to egg eating than are gulls), are constantly being driven away from nesting areas by the "resident" gulls. Thus, while eiders lose some eggs to the breeding gulls, they likely would lose a lot more if transient gulls and crows were not kept on the move.

From two to seven eggs, averaging about four, make up the eider's clutch. The eggs are protected by the female's thick layer of down feathers. When the females leave their nests voluntarily, they cover the eggs with down to maintain proper temperatures and to camouflage them against predators. But once incubation starts, females leave their nests only occasionally for drinking water—perhaps two or three times a *week*. They do little or no feeding during the twenty-six days required for the eggs to hatch, utilizing body fat and protein stored up in the weeks prior to nesting. This apparently is the way eiders have adapted to colonial living while surrounded by their natural enemies.

Meanwhile, the drakes lose interest in their mates as soon as incubation begins. Gradually, the males congregate, and eventually a large flock is formed. They leave the nesting islands and move seaward to molt near offshore ledges and shoals. Few males are seen around the bays and inner islands after mid-June. During the summer molt, they acquire a dull, gray-brown plumage, similar to that of the females. This makes them much less conspicuous at the critical portion of the molt when they are unable to fly. Eiders, like other ducks, shed all their flight feathers at the same time. Thus, for a period of three or four weeks, until the new feathers have developed, the birds are dependent on their swimming and diving abilities to escape danger.

Eider ducklings hatch almost simultaneously, and soon after their down has dried, they are led from the nest by their mother. Usually, this is within 24 hours of hatching; and once away from the nest, the birds do not return. The journey from the nest to the water is one of the

most hazardous parts of the ducklings' lives, for they must run a gauntlet of hungry gulls. Many do not make it. Once on the water, the group may join other eider families, and frequently these aggregations contain females which have lost their own nests or which are non-breeding yearlings. Such groups, or "creches" as they are called, provide additional security for the ducklings from attacks by gulls or other predators. It takes at least two months for the young eiders to reach the flying stage.

MAINE IS THE ONLY locality in the United States besides Alaska that has nesting eiders, and Alaska birds belong to a different subspecies. Thus, any steps to safeguard the U. S. breeding population of American eiders must be done in this state. Moreover, Maine and Massachusetts combined make up the principal winter range of the American eider—not only of Maine-raised birds but many of those that were reared in Quebec, Newfoundland, Nova Scotia, and New Brunswick. In addition, two other kinds of eiders, the northern and the king eider, come to the Maine coast in small numbers during winter from Arctic and sub-Arc-

Eiders often make little effort to conceal their nests. The usual number of eggs laid is four.

Photo by the author

tic regions of Labrador and Baffin Island.

For the past 11 years, a major project of the Maine Cooperative Wildlife Research Unit has dealt with the eider. Basic studies on breeding biology and productivity have been conducted on two study areas in Penobscot Bay. The Wildlife Unit and the Division of Law Enforcement, U. S. Fish and Wildlife Service, have jointly carried out a statewide investigation of population trends and a classification of the nesting islands.

The history of the eider in Maine presents a refreshing contrast to that of many other North American ducks, some of which are currently showing declines. Nesting eiders along our coast have apparently doubled their numbers at least four times in the past 30 years or so and have also extended their range southwestward.

As judged from published accounts, a gradual decline in eider numbers took place from the mid-1800's to shortly after the turn of the 20th century. According to Gross (1944), the low point of breeding eiders in Maine was reached in 1907. Following a summer survey of the Maine

Photo by Richard Ferren for Maine Cooperative Wildlife Research Unit

A female eider on her nest in a clump of weeds and shrubs.

coast, Norton (1907) wrote that probably only one breeding site remained occupied, Old Man Island in eastern Maine. Practically all the early writers said that the chief causes of the decline were over-shooting at concentration points, especially since spring gunning on breeding islands was then legal, plus collection of large numbers of eggs as food.

Abolition of spring shooting, several years of complete closing of the eider hunting season, and the lease and patrol of several important seabird breeding islands by the National Audubon Society then followed the decline. This latter move was not aimed exclusively at eider protection, for herring gulls, common terns, Arctic terns, and double-crested cormorants had also shown serious population decreases.

By 1915, these colonial breeders, including the eiders, were staging a comeback. Except for the terns, this trend has continued to the present. The population increase was gradual at first, but since the late 1930's, it has been accelerated. In addition, great black-backed gulls, that formerly nested primarily in Canada, are now increasing rapidly in Maine.

The striking change in the current status of the eider is clearly evident in considering the findings of Gross (1944). Up to 1943, Gross had recorded breeding eiders on 31 islands and probable nesting on 14 others. He estimated the state's population as "probably" more than 2,000 pairs. Less than 30 years later, 1970, we had found eiders nesting on more than 150 islands; and we estimated, from our spring aerial inventory, that the Maine breeding population was a minimum of 20,000 pairs. Also, eiders were

nesting much further southwestward than previously recorded, on several islands in Saco Bay of York County.

The rapid rise in the eider population apparently has leveled off during the past five years. An aerial census in 1972 indicated about the same number of breeding pairs as in 1970. At present, by our aerial survey of 1974 and ground checks of 1975, it appears that we still have in the vicinity of 20,000 pairs. Approximately 60 per cent of the breeding birds are in the midcoast region, roughly from Isle au Haut to Pemaquid.

EVEN THOUGH the population now seems more or less static, it "leveled" at a reasonably high point. So one might ask: why the concern for Maine eiders and their nesting islands? We should not be complacent. There appear to be several reasons to be seriously concerned for the future of the birds.

There is no evidence at present that the eider is being over-harvested. But interest in sea duck hunting is increasing. Since the eider is the only sea duck that breeds in Maine, we have a strong incentive both to preserve sufficient nesting habitat and, through wise regulations, to guard against too great an increase in the harvest.

The eiders' habit of concentrated nesting and the fact that they share breeding space with herring gulls and black-backed gulls present problems. The loss of eider eggs and young has already been mentioned. In addition, high nesting densities may be factors in the spread of epidemic diseases and parasite infes-

tations. Already, we have had three epidemics of fowl cholera, a serious bird disease, among eiders in the mid-coast region—in 1963, 1970, and 1974. The 1970 outbreak was especially widespread, and we estimated a loss of up to 20 per cent of the nesting female eiders among several of the important colonies of Penobscot and Muscongus bays and the outer islands.

However, our studies indicate that Maine eiders probably can recover quite readily from most "normal" disease outbreaks. Also, the eider usually can successfully nest in the midst of a gull colony if left to itself and not disturbed by humans. If this were not so, it would be hard to explain the expanding populations of gulls and eiders that occurred simultaneously.

But the influences of man and the increasing use of the marine resources may be too much for eiders to cope with successfully unless we effect more safeguards against disturbance to nesting birds and insure habitat preservation on the most important breeding islands. Coastal development is by no means confined to commerce and industry. Recreational development, whether by agencies or by private interests, is a matter of increasing concern. The present trends in outdoor recreation often are not conducive to successful nesting by eiders and other colonial seabirds. The current boating boom has resulted in unprecedented demands for launching sites, marinas, island picnic sites and camping facilities, and other environmental alterations.

And conservationists themselves often unintentionally add

Photo by the author

Typical eider nesting site consisting of low growth of grass and herbs on a small rocky island.

Hunting eiders on an offshore ledge in late autumn can afford exciting but cold sport for Maine's coastal duck hunters.

to the problems of the birds. The nature enthusiasts who roam through a seabird colony to look and to photograph may cause extensive losses of eider eggs or newly-hatched ducklings by predatory gulls in a matter of minutes. Observations that we have made from blinds set up in eider colonies have demonstrated what happens. When people land on an island and start to travel over it, gulls and eiders alike leave their nests. But invariably, upon departure of the visitors, the gulls return much sooner than the eiders. Heavy destruction of exposed eider eggs results.

IN CONTRAST to the situation affecting many other wildlife species, management of Maine's eider population is relatively simple. The primary needs are preservation of breeding habitat, and freedom to the birds from human disturbance during the nesting season, from May 1 to mid-July. For the remainder of the year, use of the islands for hunting, picnicking, or sight-seeing could be permitted.

In 1969, following our aerial and ground surveys of Maine's islands, about 50 of those used by eiders for nesting were classified as being of major importance. These were given a priority rating based on colony size, growth potential, and vulnerability to ex-

ploitation. These islands are well scattered from Washington to York counties. The priority list has been revised occasionally to meet changing conditions and is available to conservation agencies, both public and private, for their use in acquiring, or preserving through agreements, the habitat needed for successful breeding by eiders and their seabird associates.

Public ownership is, of course, the best guarantee for the long-term future. But with increasing land costs and with austerity budgets of most agencies these days, purchase of islands is lagging. However, restrictive use policies and regulations such as provided by the Department of Environmental Protection, Land Use Regulation Commission, and Bureau of Public Lands are a second-choice alternative. In this regard, the conservation easement program of the Maine Coast Heritage Trust has promising possibilities.

Can we justify taking coastal islands "out of circulation" for the benefit of birds? It would seem that this should be easy. Even if all 50 priority islands were acquired, less than 3 per cent of the 1,700 vegetated islands on the Maine coast (Gross 1944) would be involved. On an area basis, the percentage would be even lower. According to Dun-

nack (1920), there are more than 400 islands that exceed 1,100 acres, and most eider islands are smaller than 10 acres. There is some uncertainty as to the number of islands along the Maine coast. Various published estimates range from 2,000 to more than 3,000, depending on one's definition of an island. The Coastal Island Registry Office of the Bureau of Public Lands lists about 3,300, but these include many unvegetated ledges — any land mass above water at high tide.

Thus, the needs of eiders for islands appear very modest indeed. Some progress in meeting the needs has been made. Nearly 40 eider breeding islands are now under agency control or ownership, as follows: Maine Department of Inland Fisheries and Wildlife—25; Nature Conservance—5; U. S. Fish and Wildlife Service—3; National Audubon Society—3; and Maine Audubon Society—2. This is certainly a good start, even though only 7 of these are from the priority list. We hope a few more islands of major importance can be purchased by wildlife agencies and thus help insure that present eider populations can be maintained. ∎

Literature Cited

Dunnack, Henry E. 1920. The Maine book. Maine State Library, Augusta. 338 pp.

Gross, Alfred O. 1944. The present status of the American eider on the Maine coast. The Wilson Bull., 56(1):15-26.

Norton, Arthur H. 1907. Report of Arthur H. Norton on colonies of birds in Maine receiving special protection in 1907. Bird-Lore, 9(6): 319-327.

Maine's Dabbling Ducks

By J. William Peppard

WATERFOWL throughout Maine, the North American continent, and all over the world for that matter, are divided into two major groups — dabbling ducks and diving ducks. This article is concerned specifically with those species of dabbling ducks that are native to Maine and known to all Maine duck hunters.

Dabblers and divers obtained these names from their respective habits of feeding. Dabblers more commonly feed on plants growing out of the water or at or near the surface of the water, and while they may "bottoms up" to reach down, they seldom disappear under the surface. On the other hand, the divers usually dive under the surface to varying depths in order to reach their main food supplies. There are many other differences between these two groups of ducks, and anyone truly interested in ducks or duck hunting should consider obtaining a copy of *The Ducks, Geese and Swans of North America* by Francis H. Kortright.

The most common dabblers in Maine in the order of their abundance are the black duck, wood duck, blue-winged teal, and the green-winged teal. Occasionally, some other species of dabblers nest here, but their presence is

rare, and they are not considered to be residents. Likewise, during the spring and fall migrations, many other species of dabblers are observed in Maine as they pass through here on their way south or north.

Starting in the early spring, and as soon as the fresh-water marshes commence to open up, the black ducks are the first to return to their traditional nesting grounds. While they are closely followed by the woodies and eventually the teal, the blacks are considered the real harbingers of

spring by many Maine waterfowlers. Naturally, with such a varied climate, the return of the Maine dabblers to their nesting grounds must be gradual during March, April, and May and "in tune" with the spring thaw. This is essential to survival, in order to avoid extremes in cold weather and to insure that food and nest-

Dabbling ducks spring into the air on take off (left) while diving ducks patter along the surface for some distance before becoming airborne.

Sketches by Alan R. Munro

16

ing cover plants are also on schedule.

While many of the dabblers arrive home all "paired up," the rest will be in small flocks of three or more, with the males trying to win the females by using courtship antics displayed both on the water and in the air. Eventually, one male wins the female, the pair immediately establish their "territory" on the

wood duck which nests in tree cavities or man-made nest boxes located near the water. However, a variety of situations will suit, and black ducks, in particular, may nest in shrubs right next to the water or under a brush pile one-half mile away from the marsh. Generally, the teal prefer more grassy cover on the edge of a field or meadow handy to the water.

site. During these daily visits, the female also improves the nest and its protective cover; and finally, on completion of the clutch, which will contain eight to fifteen eggs, the nest will also be completed. In addition, the female will have plucked a complete blanket of special down from her breast, to protect the eggs from being seen when she is absent and to help maintain a near constant temperature during the forthcoming incubation period.

The incubation period starts only after the last egg is laid; it lasts on an average of twenty-eight days for the Maine dabblers. During this period, the

Legs of dabbling ducks (left) are placed near center of body, but on divers they are set near rear of body.

marsh or flowage, and any intruders of the same species will be driven from that area. This is why one will seldom, if ever, see more than one pair of any species on a small flowage in the nesting season, but one pair of each species may be present on that same marsh.

Once the territory is established, the next step is for the female to choose her nest site, which may or may not be within the established territory. All the dabblers in Maine are ground nesters usually, except for the

As SOON as the nest site has been chosen, the female immediately starts nest construction and egg-laying almost simultaneously. Building materials include leaves, twigs, grasses, and a feather now and then to start with. During this period, she remains with the male on their territory for most of the day and leaves him only for an hour or so in order to visit the nest and to deposit one egg. The male usually awaits her return on the territory, but he sometimes accompanies her in flight over the nest

female remains almost constantly on the nest, making only infrequent trips to feed with the male on their territory. The female must be extremely careful of predators both on the ground and in the air; this means that she must leave and arrive at her nest as secretly as possible, to avoid observation or detection. And once on the nest, she must remain as still and quiet as her incubation duties will allow. Nest predation is a fact of life in the wild, and the mother duck knows it, so she takes every precaution possible to avoid attracting crows, raccoons, mink, and other possible predators to the vicinity of her nest.

Meanwhile, back on the territory, the male is becoming more and more disenchanted with his mate, due to the lack of attention; finally, as the incubation period draws to a close, the male gives up and goes off in disgust looking for other males of the same species with similar problems. Gradually, these disgruntled males flock together, and by early

Dabblers (left) tip up to feed, but divers dive completely under water.

summer they have entered their post-nuptial "eclipse" moult which lasts about one month. They usually choose a very densely vegetated area to avoid observation, because during this period, they lose their beautiful breeding plumages including their flight feathers. For a short time, the resplendent drakes resemble their females in coloration and can't even fly! However, as the summer wears on, they gradually regain their flight feathers and begin to show up in areas where they can be observed. These same males moult once again in the early fall, but this time they do not lose their flight feathers, and their new plumage begins to resemble the brilliant breeding plumage which they lost earlier.

INCUBATION ceases when the first egg begins to "pip" or crack, and within twelve to twenty-four hours all of the eggs will have hatched, because incubation did not start until all eggs were laid. After a few hours to dry off and to get squared around, the mother will lead her young to the marsh and never return to the nest again. Young wood ducks, though, have a unique problem. Their nest may be fifty feet in the air in a hole in a tree, their mother is on the ground calling, and they don't know how to fly! However, they make up for this by being very light and fluffy, and when mother calls, they simply "scramble" out of the cavity and sail down to the ground, with very little grace but lots of buoyancy!

Once on the marsh, the brood period begins in earnest, and while mother duck does not have to teach the young how to swim, she does have to teach them how, when, where, and what to eat, and how to protect themselves against all the different kinds of dangers they will be encountering. Also, father duck is not available for help, because he is busy moulting somewhere else, so all the responsibility for survival

rests on mother duck during the brood period.

During the early stages of the brood season, the young dabblers feed mainly on insects found both on plants and in or near the surface of the water. Gradually, they learn which fruits or seeds to eat and also which roots or tubers under the water's surface are good for them.

As the weeks go by, the young become stronger, get more active, and eventually start trying out their wings. While actual flight age varies between species, and

HIND TOE NOT LOBED

HIND TOE LOBED, FOOT LARGE

The hind toe of dabbling ducks (top) is not lobed, and the foot is smaller than in diving ducks. Larger feet and lobed hind toe of diving ducks (bottom) are an aid in swimming and diving.

even within a brood, most young dabblers in Maine are usually on the wing sometime between six and eight weeks of age. Once flying, the young ducks begin to disperse from the home marsh; banding records show that the young birds of a given brood may leave in several different directions even though, eventually, all of them will fly south for the winter.

As the brood disperses, the responsibilities of the female end,

and she enters a moult and flightless period comparable to that experienced earlier by the male. When she emerges from this moult in early fall, she has a bright new winter coat for the coming cold weather.

AS THE summer draws to a close, the Maine dabblers begin to flock together for the long flight south to their respective wintering grounds. The blue-winged teal usually start first and are followed in order by the woodies, green-wings, and the black ducks. The hunting season in early October probably hurries the southward movement somewhat, so that by early November only the hardy black ducks remain in Maine in any numbers. Again, banding records indicate that most Maine green-wings spend the winter months in the Gulf states and Central America, and the blue-wings in much the same area except a little more south to South America and the West Indies. The wood ducks also winter along the Gulf coast and the southern Atlantic coastal states. And while the black duck winters throughout the south, concentrations of blacks may be found in all of the Atlantic coastal states from Maine to Florida.

After what must be a harrowing trip south through many different states with different hunting seasons, all of these species are content to remain quite inactive on their respective traditional wintering grounds for a few months. However, the life cycle must go on; as the breeding urge returns in late winter, the Maine dabblers start their northward trek back to their home nesting grounds. And while we don't know exactly where they've been or how they've managed, we do know that a pair of blacks, woodies, or teal swinging their marsh in Maine in early March or April is a sure sign that winter is about over, and spring is on the way! ■

Maine's Diving Ducks

By J. William Peppard

© Leonard Lee Rue

IN THE 1967 spring issue of **Maine Fish and Game**, we presented brief life histories of the native fresh-water dabbling ducks which are residents of Maine. We will now discuss the other major group of ducks, the divers.

As their name implies, divers search for their food by swimming beneath the surface of the water. Other important differences between these two major groups of wildfowl as listed by Francis H. Kortright in *The Ducks, Geese and Swans of North America* include the following: (1) divers tend to frequent more open and deeper water areas than do the dabblers; (2) divers require more water surface for landings and take-offs than do the dabblers who have the ability literally to "jump" from the water when necessary; (3) the colored wing patch or speculum on the divers is usually less bright than on most dabblers; (4) the hind toe on the divers has a pronounced lobe which is barely noticeable on the dabbling ducks. All of these differences are easily recognized by experienced duck hunters and are readily learned by inexperienced hunters.

In Maine, the four most common resident species of fresh-water diving ducks are the ring-necked duck which is a comparatively new nesting species in Maine, the little hooded merganser or "humpback," the large American merganser or "shelldrake," and the American goldeneye or "whistler" as it is often called. And as implied by the names, this group may also be split into common diving fowl (ring-necks and goldeneyes) and the mergansers (hooded and American). The latter group may easily be identified by the long, narrow, tooth-like bill as compared to the flat duck bill on the other divers. The bill on the mergansers has been especially developed for catching and feeding on small fish, tadpoles, and other slippery prey. Consequently, they are sometimes called "sawbills" or "fish ducks" in Maine. Their food habits are often reflected in the odor of their cooked flesh, and many a beginning duck hunter, or rather, his wife, has been surprised and disappointed to learn that the trophy in the oven smells more like fish than fowl! However, many people in Maine, especially along the coast, really enjoy boiled potatoes and onions in a hot steaming "shelldrake stew" on a cold winter's evening.

IN THE SPRING, the divers return to Maine from their respective wintering grounds once again to nest and to rear their young. The goldeneye and the American merganser are always the earliest arrivals, and they are followed by the hooded merganser and the ring necked duck. While goldeneye nesting is pretty much limited to northern Maine, the American merganser nests throughout the state. These two species are also well known for their courtship displays which are readily observed during March or April. The males of both species are real show-offs, and a courting party usually consists of five or six males competing for the attentions of one female.

The displays include short erratic chases, both on the water and in the air. The most spectacular display occurs when one male points his bill towards the heavens, then arches his head back to touch his tail, and with a quick snap suddenly rocks forward and, in so doing, kicks out a splash of water to the rear. This "head-bobbing" by one male stimulates the others to go through the same performance, which makes the courting party a pretty active group except for one! The female, passive and apparently unconcerned, simply drifts around and acts as though she could care less! The hooded merganser is much more secretive during courtship, and the ring-neck is less colorful; but, eventually, pairs are established for all species, and so are their respective "territories" and nesting sites.

The goldeneye and the hooded merganser prefer to nest in cavities in trees, but both will utilize wooden nesting boxes when they are available. The American merganser apparently also prefers high nest sites such as cavities in trees or cliffs, but it often nests on the ground. The ring-necked duck always chooses a nest site on the edge of the marsh so that she can swim directly to it. Unfortunately, this particular habit of the ring-necks makes many of their nests vulnerable to flooding after heavy rains.

The number of eggs per nest for all species concerned will vary from five to fifteen but usually averages between eight and twelve eggs. Incubation of the eggs varies from three to four weeks, and this is completed solely by the female of each species. Toward the end of the incubation period, the males depart from the area in order to enter their "eclipse moult" which reduces their dressy breeding plumage to a more drab plumage similar to that of the female. Like all other ducks during this particular moult, they are flightless for a period of time until their new feathers are re-established and strong enough to permit flight.

Since all of the eggs in each nest hatch on the same day, the broods are usually ready to leave the nest for good by the second day. The young of the goldeneye and the hooded merganser may be forty feet up in a hole in a tree, and they have the same problem as young wood ducks. However, a few soft calls from the mother on the ground will usually entice the young out of the hole and down to her side. Naturally, the bravest go first, but within a few minutes all have made their first "flight" and are safely with their mother. The ground or marsh-nesting species do not have this problem; the young ring-necks simply scramble out of the nest onto the water and immediately know how to swim.

The brood season lasts throughout the next couple of months, and during this period the young have much to learn about survival. In fact, "survival of the fittest" is the watchword all through the brood season. Many dangers threaten the brood constantly, and an average brood of ten or twelve at hatching will do well

to have five or six left by the time the flying stage is reached. In addition, the young must also learn about foods and eventually about flying, which often appears to be a case of trial and error!

As FALL approaches, the ring-necks and hooded mergansers are the first divers to start their migration towards their wintering grounds in the southern Atlantic and Gulf states. The goldeneye and the American merganser appear to be more hardy, and besides migrating later in the fall, neither do they go as far south; many of both species winter in the bays and estuaries on the coast of Maine. And while the early migrants provide some shooting during the first part of the duck season in October, "whistler" or goldeneye shooting usually starts later on, in November; the first arrivals are eagerly awaited by many hardy duck hunters on both the inland rivers and lakes and along the coast.

Normally, throughout the winter months, all of the divers are busy trying to survive the *natural* hazards including migration, predation, and weather. Today, however, like all other species of waterfowl, they also have to contend with many *unnatural* obstacles to the completion of their life cycle. Some of these obstacles include hunting seasons, pollution, and marsh drainage and/or development which often jeopardize their traditional wintering grounds.

However, all four species of divers in Maine are apparently holding their own, thanks to both modern day law enforcement and wildlife management practices and recommendations. In late winter, the goldeneye and the American merganser are among the very first of all ducks to move inland. And on a real cool evening in early spring, the whistling wings of goldeneyes overhead tell many folks in Maine that even though the lakes are still frozen, there is a river or stream open nearby, and the hardy whistlers are headed north for another nesting season. ∎

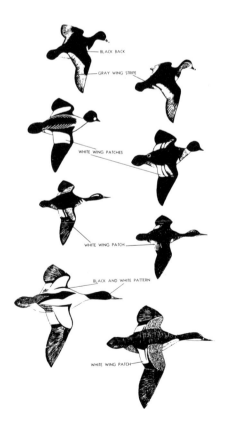

BLACK BACK

GRAY WING STRIPE

WHITE WING PATCHES

WHITE WING PATCH

BLACK AND WHITE PATTERN

WHITE WING PATCH

Key identification characteristics are noted, top to bottom, on: ring-necked duck, American goldeneye, hooded merganser and American merganser. Females of each species are on right.

Sketches by Alan R. Munro

Female (right) and male ring-necked
ducks in spring plumage.
(From a painting by Ralph S. Palmer.)

Maine's Duck of Many Names —

The Ring-neck

By Howard L. Mendall

I F YOU WERE hunting ducks last October on one of Maine's inland marshes, ponds, or deadwater streams, the chances are pretty good that you shot, shot at, or at least saw a ring-necked duck. And if you are an average hunter who gets afield only a few times each fall, the chances are also pretty good that you didn't recognize the ring-neck as such.

To carry our "chances are" story a little further, your encounter with ring-necks could easily have gone like this: You were crouched in a blind with a spread of black duck decoys in front. But the blacks weren't flying well on this "bluebird" morning, and you got restless and inattentive. Maybe you started to open your lunch. Suddenly, there was a swishing rush of air behind you—half whistle and half roar, like jet planes in the distance. Six or eight medium-sized ducks swooped over your head,

circled the decoys, flared, and swung out of range before you could get your gun to your shoulder.

That was likely a flock of ring-necked ducks, one of the speediest of waterfowl. This is the species that Allan Brooks, internationally famous bird artist and a student of flight speeds, rated as the second fastest duck in North America— exceeded only by the old squaw. Some Maine duck hunters are inclined to dispute Mr. Brooks. They would put the ring-neck first.

If you do not recognize this duck, you are not alone. Many hunters have difficulty in separating ring-necks from several other species, especially the lesser scaup or bluebill. The two ducks are similar in size, often occur on the same water-

ways, and have a rather close resemblance. Careful scrutiny, however, reveals several differences. The back of the male ring-neck is completely black, in contrast to the scaup's whitish back with only flecks of black. The ring-neck's bill has a white transverse band and, in winter and spring plumage, is outlined in white. By contrast, the bill of the scaup is pale blue without the white markings— hence, its popular name of bluebill. When a drake ring-neck is seen on the water, the white plumage of the underparts extends in front of the folded wing to form a conspicuous crescent which no other black and white duck has. In size, it is similar to our common goldeneye or whistler.

The female ring-neck is not so distinctively marked. She is mainly brownish-gray with mottled white underparts that become brown in summer. Her bill has the white, transverse stripe of the male, but it is less prominent. Both sexes have a gray wing patch, and when the bird is in flight, this will separate it from scaups or any other common, medium-sized duck in Maine.

The chestnut neck-ring of the male, which gave the duck its name, is seldom seen except when the bird is viewed under favorable light conditions. Thus, this is not an appropriately named duck. It would seen more logical to call it the ring-*billed* duck. In fact, over much of the northern part of its range, this *is* the name by which the bird is known to most sportsmen. But the ring-neck has many other popular names, depending on locality. Some of these are: marsh bluebill, little bluebill, ring-billed bluebill, marsh broadbill, little broadbill, ring-necked scaup, ring-billed scaup, moonbill, little raft duck, and blackhead. A name widely applied throughout the South is blackjack.

The female ring-neck is a rather noisy duck, especially during spring and when she is disturbed with her young. Her call can best be described as a "purring-growl." The male is usually silent except during courtship display, when he utters a low, hissing whistle audible for only a short distance.

THE RING-NECKED DUCK is not a "native" of Maine, or anywhere else in the Northeast, for that matter. It is a comparative newcomer, a bird that has made a striking change in distribution. Many birds have extended their ranges, and others have undergone drastic numerical fluctuations. But it is doubtful that any of these changes occurred more rapidly over such a large geographical area than the eastward journey of the ring-neck. Formerly, it was a breeding bird of the West and Midwest. In the New England states and Atlantic provinces of Canada, it didn't even occur in migration except on an accidental basis.

An abrupt change in the ring-neck's summer quarters took place during the 1930's. The bird became increasingly common in the Northeast during migration. It was found breeding in Maine and Pennsylvania in 1936 and in New Brunswick the following year. Then, within a few years, came new nesting records for Nova Scotia, Prince Edward Island, Quebec, Newfoundland, eastern Ontario, New Hampshire, Vermont, New York, and Massachusetts.

Not all of these new breeding nuclei were successful, but many were, and the population increased and spread out still further. The duck is now well established as a breeder throughout much of the Northeast. It is the second most abundant inland nesting duck in Maine, outnumbered only by the black duck.

The ring-neck, a diving duck, obtains most of its food beneath the surface. It consumes more vegetable food than do most divers. The seeds and tubers of bulrushes, pondweeds, and burreeds make up a major portion of its food in Maine. Other important items include seeds of sedges, smartweeds, water lilies, and wild rice; also rootstalks, buds, and leaves of wild celery. The downy young depend heavily, as do most young waterfowl, on aquatic insects, small snails, and other animal foods rich in proteins.

Ring-necks are primarily birds of fresh-water marshes and are seldom seen on the coast. Although they are often found in tidal rivers and estuaries, it is usually in the fresh-water portions of such habitat. They are very partial to sedge-meadow marshes and bogs that are numerous in northern, eastern, and central Maine. This type of habitat commonly occurs in ponds, the coves of some of the large lakes, and, especially, thoroughfares between lakes. Sluggish woodland streams and reed-bordered dead-waters of rivers are likewise favorite environments.

During recent years, ring-necks in both Maine and New Brunswick have made increased use of beaver flowages for nesting. It is the largest and older beaver ponds—those with grass and sedge borders or with numerous sedge hummocks—that the ducks seek. When the birds first "moved east," they seemed to shun beaver ponds and were almost exclusively confined to sedge-meadows and bogs. Perhaps they have learned, through their association with black ducks, that beaver ponds—with rather stable water levels and freedom from human disturbance during the breeding season—make fairly safe nesting habitat.

This central Maine bog pond is typical ring-neck nesting habitat.

Photos by the author

The canoe paddle marks the site of a ring-neck nest that was used for four consecutive years. A tunneled runway, shown by the arrow, led to the nest.

has appeared and when there is less danger of spring floods.

The nest site is selected by the female, but she is accompanied by her mate. At times, the preliminary selection is from the air, with the pair making low passes over the marsh. Always, whether by flying or swimming, these searches are by the female leading and the male following closely behind.

When breeding is successful, the birds invariably return to the same nesting area—often to the identical site. One nest on a floating island at Portage Lake was used for four consecutive seasons. At the state's Pennamaquan game management area, a bird live-trapped on her nest was first banded as a duckling two years before, only a hundred yards from where she was found nesting.

The nest may be in a tussock of grass or sedge, sometimes with no additional cover but more usually associated with a clump of low shrubs. More nests in Maine have been found in a combination of

sedge-sweetgale-leatherleaf than any other cover type. The nest base may be virtually surrounded by water, whether in a wet marsh or on a small island. Floating islands of northern bog lakes and ponds, and the floating sedge mats that border many woodland streams, are favorite nest sites. Their wet situations discourage many of the land predators, and their "float-ability," while not unlimited, does allow for minor fluctuations in water levels without danger to the eggs.

The nest is quite shallow, constructed only of grasses and sedges that are within reach of the female as she sits or stands at the site. She lays one egg a day until the clutch, which averages nine, is complete. Toward the latter part of the laying period, the hen gradually adds a nest lining of breast down to afford insulation for the eggs. When she leaves the nest to join her mate and to feed, she covers the eggs with the down, thus making them less conspicuous to a potential predator.

A ring-neck's nest with 10 eggs in a clump of sweetgale and sedge. Vegetation in the foreground was pulled aside to permit the photograph.

THIS DUCK is not a cold weather bird. Although it arrives in Maine from the wintering grounds of the southern states in April, the peak of migration may not be reached until the end of that month. Some black ducks are already nesting while ring-necks are still leisurely traveling northward.

Even after the birds have reached the breeding marshes, they seem to be in no hurry to set up housekeeping. Often, they linger for two or three weeks in loose flocks, in the open water of the breeding marshes or on a nearby river or lake, before individual pairs slip away from the group to select nest sites. This delayed nesting may be nature's way of insuring better hatching success. The ring-neck is very exacting in its nest site—it must be a fairly dry spot but as close as possible to swimable water. These requirements can best be met by waiting until well after the spring run-off when new marsh vegetation

Photo by the author

Although only a few hours
old, these ring-necks
are nearly ready to leave the nest.

INCUBATION BEGINS when the last egg is laid and continues for 26 or 27 days. Generally, all young hatch within a span of 5 or 6 hours, and in another 2 to 4 hours, the down of the ducklings is dry. The amount of time the young are brooded in the nest depends on the weather and the time of day hatching occurs. Ducklings that hatch during the night or in early morning are usually led from the nest by the female before sunset. If, however, hatching takes place during mid-day or afternoon, they remain in the nest overnight. Cold, wet weather usually delays nest departure, while disturbance by predators or humans hastens it.

The ring-neck is one of the most devoted of all duck mothers. Her "broken wing" or injury-feigning act in time of danger is much more persistent than that of many other species and often appears quite effective in encounters with some of the mammalian predators such as minks and foxes. Ring-neck hens generally remain with their broods throughout the entire rearing period, in contrast to many ducks which often desert the young when the latter are half or two-thirds grown.

During their first few days, the downy young, which are a deeper yellow than any other Maine duck, feed on insects found on the surface or among emergent vegetation. But when less than a week old, they start making short dives for soft-bodied animal life beneath the surface. By the time the ducklings are half grown, they are skilled at diving and submerged swimming. The first flights are made at seven-eight weeks of age, and then the young are on their own. They may remain near the breeding marsh until the southward migration, or they may depart for larger lakes or rivers to join other ring-necks.

Bringing her family of ducklings to the flying age is not a simple task for the hen ring-neck, who faces many adversities. Despite the duck's tendency to locate the nest on a semi-floating base, heavy rains sometimes cause water levels to rise and flood the eggs. Predators of eggs are a constant threat, especially crows, ravens, minks, raccoons, and foxes. The last three occasionally kill incubating females as well as eat eggs and ducklings.

An important factor influencing nesting success is man. Most ducks, but especially the ring-neck, desire peace and quiet on the breeding grounds. Frequently, disturbances by picnic parties, sightseers, and especially traffic by boats with high-powered outboard motors, cause nest desertions, induce more predation than normal, and disperse the broods.

All too often, some of the ducklings that become scattered never rejoin the family, and they soon die.

THE RING-NECK is not an important game bird in Maine as a whole and makes up less than 5 per cent of the annual waterfowl harvest. However, on a local basis, it is often of considerable importance. This is especially true during October hunting in the interior of Washington County, in the bog lakes of Aroostook County, and in portions of the Mattawamkeag, Penobscot, and Sebasticook drainages and in the Saco Valley. On some of these marshes, early season hunters may shoot as many ring-necks as black ducks.

What caused a portion of the mid-western population of ring-necks to make such a remarkable change in breeding range is an unanswered question. That it was not a gradual eastward movement but an abrupt jump over many hundreds of miles makes it even more noteworthy. Was a sizeable flock in northward migration forced far off course by a major weather disturbance and found themselves at breeding time in waterways that resembled their normal range? Did extensive habitat loss during the famous drought of the 1930's necessitate a deliberate search by the birds for new habitat? Some evidence would support this theory because of the timing. The mid-western droughts, dust bowls, and loss of waterfowl habitat did coincide, as best we can determine, with the arrival of the ring-necks in Maine and New Brunswick. Whatever the reason, the success of the "transplant" undoubtedly lies in the fact that the ring-neck has the biological qualities of a pioneering species. Otherwise, the scattered little colonies of ring-necks would have existed only briefly, then faded into history.

SO, WHETHER THEY call him a ring-neck, ring-bill, marsh duck, little bluebill, or blackjack, the duck hunters and bird watchers of the Northeast are grateful for this transplanted mid-westerner who is now well established a thousand miles from where nature intended him to be. ∎

Ducks That Nest in Maine

By Howard E. Spencer, Jr

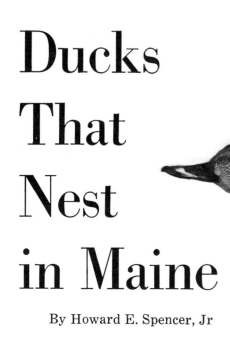

The black duck, Maine's most common breeding duck, is identified readily in flight by the silvery underwing linings.

MOST PEOPLE have no difficulty recognizing a brood of ducks when they see them around our lakes and rivers in June and July. Surprisingly few, however, are aware that these little bits of fluff may be any of ten or eleven different kinds of ducks that nest in Maine. This article will be a sort of bird-bookish attempt to help the reader learn to recognize these various species and provide a smattering of information as to what role they play in Maine's waterfowl picture.

Quite honestly, identifying the female duck and her brood is not the easiest thing in the bird-watcher world, and it may take considerable experience before one becomes confident in saying "that's a brood of ringnecks" rather than green-winged teal. The gaudy colored males are rarely, if ever, present to aid the observer, and the almost universally drab plumage of the females often shows poorly, at best, in the marshy vegetation, as she hurries her young into hiding. Even her flight behavior changes as she simulates injury to distract the pursuer. Instead of leaping almost vertically from the water as dabbling ducks are supposed to do, mother black duck may flop quite helplessly along on the surface. The fact that divers like the ring-necked duck have a lobed hind toe doesn't really help much in identifying her as she and the brood swim off into the pickerel weed.

Although there is no easy way around these problems, there are some features that will help and to which the observer should pay particular attention. For example, the color of the legs, feet, and bill are sometimes more easily distinguished than body plumages and may help if they can be seen. Wing patterns and voice may also prove helpful. Similarly, knowing what kinds of ducks are common to Maine, and the general type of cover they prefer for nesting and rearing their young, may tend to eliminate several species. In some cases, the breeding distribution within the state may rule out such species as the common goldeneye, which we have not recorded as a breeding species south of Lake Cobbosseecontee in the vicinity of Augusta.

To provide some order to this discussion, the eleven species which breed in Maine are discussed individually below in the approximate order of their nesting abundance. The common eider duck and redbreasted merganser are exceptions and will be discussed following the inland forms. As far as we know, both these latter species are largely marine, and breeding is restricted to the coastal islands.

Insight as to relative abundance of the inland nesting ducks is provided by the composition of the following 371 identified broods tallied during production studies made by the Game Division in 1965 and 1966:

Species	Per Cent of Broods
Black Duck	37
Ring-necked Duck	21
Wood Duck	14
Goldeneye	14
Hooded Merganser	8
American Merganser	2
Blue-winged Teal	2
Green-winged Teal	1
Mallard	1
	100

The reader is cautioned to consider these figures in general terms only, since species like black ducks and wood ducks are much more difficult to observe than ringnecks or goldeneyes, and a smaller proportion of the actual population is seen.

BLACK DUCK and MALLARD: The black duck (*Anas rubripes*) is far and away our commonest and most widely distributed breeding duck. First cousin to the mallard (*Anas platyrhynchos*), it is essentially the same size, and both sexes in all plumages look alike. The head, body, and tail are largely mottled brown, and the most distinctive feature is the *silvery underwing linings* which flash clearly during flight. The black is the largest of our inland breeders except for the mallard and possibly the American merganser. It is perhaps the most likely to be seen even though it tends to be very wary in its habits. Usually, the adult female will not be confused with species other than the mallard, which is somewhat lighter in body color and often very pale around the margins of the tail. Since the mallard is a very uncommon breeder in Maine, there is but a slight chance of seeing a brood compared to the black duck. *Voice* will also help identify the blacks and mallards. The females of these two species are the *only* breeding waterfowl in Maine that make the loud pronounced "quack" so commonly known to everyone.

The downy young of most waterfowl are hard to identify in the field, and black ducks are no exceptions. The best bet is to make a serious effort to identify the mother in all cases. However, specimens and good color plates such as shown by Kortright (1943) may help. During the first two weeks of age, young blacks present a dark brown and yellow, downy appearance. Between two and eight weeks, when they take wing, the everchanging plumage is an almost uniform nondescript brown. The nearly black single stripe extending almost from the bill through the eye to the back of the skull is more pronounced than in all but the young mallards and possibly blue-winged teal. This character is of little value in the field, however.

Black ducks nest in about every conceivable wetland type and occasionally far from water. Seldom do many pairs nest closely together, and three to five broods seem to be about the most that are ever observed in any one wetland or waterway. Maintaining a maximum number of suitable breeding areas rather than maximum acreage is the best way to produce black ducks in Maine.

WOOD DUCK: The wood duck (*Aix sponsa*) is perhaps our second commonest nesting duck despite its third place rating in the table above. It is a dabbling duck, a fine game and table bird, and the male is the gaudiest of all our waterfowl. A hole-nesting species, it utilizes natural cavities in trees as nest sites and takes readily to artificial nest boxes. This species prefers the smaller, more densely vegetated or even wooded waterways. Many broods are raised on beaver flowages. Secretive and hard to observe with a brood, the middle-sized female has three main features which may aid in identifying it. On the water (and with binoculars) the rather prominent white ring surrounding the eye will separate the female woody from other species. If the bird flushes away from the observer, the white trailing wing margins also show clearly. The third characteristic is the voice when startled, which has been described as "eek eek eek," each utterance being considerably drawn out. This unusual voice has given rise to the gunner's nickname of "squealer" in some parts of the country. Wood ducks become increasingly scarce nesters in the northern and western sections of Maine, particularly on the Moosehead Plateau.

RING-NECKED DUCK: The ring-necked duck (*Aytha collaris*) is another middle-sized Maine nesting duck.

A true diving fowl, it prefers more open marshes with floating-leaved vegetation and not the dense emergent cover some of the dabblers like. This species is somewhat easier to observe with their broods than the wood duck, for example, and the downy young give a rather more *yellowish* appearance. The adult female is very nondescript and in flight shows *no white* on the wings (a dull gray wing stripe is all she has). However, the voice is quite distinctive — a very guttural "harrh" "harrh" when she is alarmed with her brood. This voice is similar to the hooded merganser female's, but the latter shows white on the wings when flushed, as well as some other differences.

The ringneck provides a limited amount of early season shooting in Maine but has largely moved south by early November. An excellent table bird, they comprise a little less than 2 per cent of the Maine harvest.

COMMON GOLDENEYE: The common goldeneye or "whistler" (*Aytha clangula americana*) is, like the wood duck and hooded merganser, a hole-nesting species that utilizes natural tree cavities or man-made boxes as nest sites. Its nesting distribution is limited to the northern two-thirds of the state. Goldeneyes are diving fowl and often rear their young around our larger, rocky, cold-water lakes. Often a brood of whistlers is quite easy to approach, and the chocolate brown head of the female is fairly distinctive. The young give a definite impression of being black and white with prominent white cheek patches. The voice of the female is also distinctive, and Kortright (1943) describes it as "cur-r-rew." It is lower pitched than the ringneck and hooded merganser and considerably less in volume. The female shows prominent white wing patches when flushed.

Although many local nesting goldeneyes depart before the hunting season, good supplies of migrating birds from the north are usually available for hunters from the last week in October on through the remainder of the hunting season. Whistlers are a good game duck, being fast and decoying readily. They're apt to be variable as a table bird, however, and occasionally become strong flavored

White eye ring and white margin on trailing edge of wing identify female wood duck.

late in the season when their diet turns to nearly 100 per cent animal matter. Goldeneyes make up 5 to 6 per cent of the Maine harvest.

HOODED MERGANSER: The hooded merganser (*Mergus cucullatus*), or "humpy" as it's sometimes called in Maine, is a third Maine nesting duck that utilizes natural tree cavities or man-made boxes for nest locations. A rather smallish duck, it is distributed statewide but is rarely, if ever, very numerous. The dull cinnamon brown head with a slight crest and long thin bill, and the traces of white on the wings, plus the guttural voice similar to the ringneck, will aid in identifying the female. Broods of this species are often seen on more open waters as they search for minnows and aquatic insects along the shores. The heads of the young are also a dull brown similar to their mothers'.

The nickname "humpy" comes from the hooded's rather humpbacked appearance in flight.

AMERICAN MERGANSER: The American merganser (*Mergus merganser americanus* Cassin) is seldom classified as a game duck due to its fondness for fish and resulting strong flavor. A large bird, nearly the size of a black duck or mallard, it may nest on the ground or in tree cavities. We have a single record of one using a nest box and incubating nine of her own eggs plus six goldeneye eggs — all of which hatched, as closely as we could determine. This species often nests around large lakes and rivers where it can satisfy its voracious appetite for fish. Seldom are many broods found on the same lake. The large size, brownish head, long pointed bill, gray body, and prominent white wing patches will help identify the female. Their flight is very direct, and the entire body, neck, head, and bill are held rigidly stiff with the bill pointing straight ahead. Their wings are set far back on the body, and they rather remind one of a jet aircraft in flight. The young are little replicas of mother, and the family groups of sheldrakes, as they're sometimes called, are perhaps the best behaved and guarded of all our nesting ducks. Very seldom do the young mergansers ever get separated from mother. They can and do travel across the water by half-swimming, half-flying, much faster than two men can paddle a canoe or row a boat.

BLUE-WINGED TEAL: The petite blue-winged teal (*Anas discors*) is the next to the smallest of our nesting ducks. It prefers marshy, grassy areas to nest and rear its young. Being a 100 per cent dabbler, it's seldom found on very deep water. Though small, it's a choice table bird and is welcomed by gunners as a tricky flyer worthy of their skill. It's usually the first to leave in the fall, and most are gone by mid-October.

The small size, relative tameness, and beautiful powder blue wing *coverts* make this a fairly easy bird to identify as soon as it spreads its wings. Depending on light conditions, the blue wings may appear almost white, gray, or even dark. But there is nearly always a strong contrast, and the identifying blue is on the upper and *forward* surface of the wing rather than the rear edge as in many species. The voice is rather weak but quite distinctive. Hard to describe, it's best learned through experience.

GREEN-WINGED TEAL: The tiny green-winged teal (*Anas crecca carolinensis* Gmelin) is our smallest nesting duck. Another true dabbler, the female averages only 12 ounces in weight. Other than its very small size, the female in breeding plumage lacks any clear distinguishing features. The green iridescent speculum on the rear edge of the wing, from which the species gets its name, seldom shows prominently enough to constitute a useful field mark. Since green-wings make up only 1 per cent or less of our breeding ducks, most observers will rarely face the problem of identifying broods of green-wings. They are usually found in habitat similar to that frequented by their cousin, the blue-winged teal.

SEA DUCKS: The last two species that nest in Maine are the American eider (*Somateria mollissima dresseri* Sharpe) and the red-breasted merganser (*Mergus serrator* L.). Both of these are marine forms, and positive records of nesting red-breasted mergansers are lacking. (Females of this species cannot be separated from the American mergansers under field conditions.) However, it is known that the red-breasted remains largely on salt water, and it has been assumed in the past that broods of mergansers in the vicinity of coastal islands were the red-breasted species.

Eider ducks, famous for their insulating down, nest in colonies on coastal islands. This is a very large diving fowl, the females averaging nearly a pound heavier than female mallards or black ducks. The female is dull mottled brown, and both sexes bear a large, strong, characteristic bill. The downy young are a uniform dull black. There is little likelihood of misidentifying these birds in family groups, because of their limited distribution. We presently know of about seventy-five islands along the coast that support eider colonies.

THIS CONCLUDES a very brief resumé of ducks that nest in Maine. For further and more detailed information, the reader should consult a good reference work such as *Ducks, Geese, and Swans of North America* by F. H. Kortright, published in 1943 by the American Wildlife Institute. ∎

Young goldeneyes appear black and white with white cheek patches.

No. 827-00239 and Associates

By Howard E. Spencer, Jr.

"She'd been hypnotized by a spotlight . . ."

No. 827-00239 was late that year in recovering her flight feathers and coming out of the eclipse moult. The biologist who removed her —somewhat bedraggled—from the live trap, up on the Penobscot River below Lincoln, recorded her as an adult female black duck (*Anas rupripes*). He placed an asterisk beside her number in his field notes to indicate that she was a recovery from some banding station other than his own. Before releasing her unharmed, he observed how thin she was for mid-September and guessed correctly that she'd hatched and raised a late brood that year. Such was No. 827-00239's second close contact with man, and her chances of surviving a third were slim, indeed.

Actually, for a bird less than two years old, No. 827-00239 was fairly well travelled. She'd been hypnotized by a spotlight mounted on an airboat in her natal marsh in Quebec the previous summer. A Canadian biologist who dipped her up in a long handled net about 2:00 o'clock one August morning identified her as a bird of the year and carefully placed on her leg the metal alloy band bearing her number. Soon after that rather terrifying experience, she'd left her home marsh and headed north. At Lake St. John, she joined a group of about 35 other young black ducks; after a few days, the whole group moved north again to Lake Peribonca. Here, however, the nights soon turned cold, skimming some of the back waters and shallows with ice, and the flock moved southwest to Baie Comeau.

They were not long at Baie Comeau, though, before the hunting season opened. Driven from this area, they moved southwestward up the St. Lawrence, losing a bird or two to hunters as the flock settled to feed or toll to decoys. Soon they left the St. Lawrence, and a short southward flight took them to the rich wild rice fields of Lake Megantic. After a few days of lush feed here, the opening of the hunting season in this zone drove them south again—this time across the international border to Merrymeeting Bay in Maine. Among thousands of their kind, they fattened for more than two weeks on wild rice, pondweeds, and bulrush before another opening hunting season provided the stimulus to send them southward again in mid-October.

By this time, No. 827-00239 had learned that open water meant safety, and she'd developed the habit of accompanying the big flocks that left Merrymeeting just after daylight to rest undisturbed on the nearby ocean, returning at dusk to feed through the night. This habit may well have been the saving factor that enabled her as a young bird to escape the hunters this first year. She moved southward along the coast, over the salt marshes of Massachusetts' "North Shore" and Cape Cod, to settle finally for the winter near Tuckahoe, New Jersey, in the great Cape May marshes.

AFTER THE GUNNING stopped in New Jersey, about the end of the first week in January, an aerial survey by biologists indicated more than 70,000 black ducks in the New Jersey marshes alone. It was soon after this that a period of extreme cold froze the marsh surfaces; the salt marsh snails (*Melampus sp.*) which No. 827-00239 had come to depend on for food were beneath an impenetrable layer of ice. Although she lost nearly a third of her fall weight and scaled only 30 ounces when a thaw made the snails available again, she survived. Others that were in poorer condition to start, or carried lead shot from the hunting season, died.

After the "freeze" had broken, life was relatively routine for the wintering blacks. Occasionally another cold snap would result in a hungry day or two. Some of the birds yielded to the temptation of the bright yellow corn so enticingly displayed in funnel-mouthed wire traps. These ducks emerged shaken but uninjured at the hands of the game biologist who saw to it they carried bright, new, number-bearing bands on one leg or the other. Several of these birds really got the habit and returned to the traps day after day for corn, much to the disgust of the biologist whose cold fingers had little sympathy for the free-loaders.

By mid-February, the lengthening days and warming sun began to affect the drakes; No. 827-00239 was receiving more than casual attention from several. After several days of head-bobbing antics, short chases to drive away other would-be suitors, and other performances calculated to win a black duck's heart, No. 827-00239 accepted as a mate a big male about a year older than she. He had come into breeding condition earlier than some of the younger birds. Once this pair bond had crystallized, the two were never far apart. The big drake defended his rights both on the water and in the air, driving away all contenders who attempted to usurp his position.

As spring progressed, the flocks began breaking up, and 827-00239 drifted northeastward with her mate in company with more of her kind. Ultimately, she came to Kennebec River estuary which was still full of ice floes. Merrymeeting Bay, its bottom covered with wild rice seed, was still ice-bound, and soon a large flock pushed eastward to the Penobscot estuary, between Castine and Winterport. Here 00239 and her mate found limitless stocks of tiny clams, snails, and crustaceans on the teaming mud flats of rich alluvial silt. Even the thousands of black ducks, whistlers (goldeneyes), and bluebills (scaup) that had spent the winter here hadn't been able to consume more than a small fraction. New birds were arriving daily, and soon the river was filled with birds and drifting ice as melting snow swelled it bankfull and broke winter's hold late in March. The urge to nest was strong within No. 827-00239. She pushed up river, followed by her mate, almost as soon as the ice was out. She made daily flights away from the main river, but everything was still ice and snow in the back country, and they'd return each night to feed and rest.

It was during this period in late April that she lost her first mate. He landed on a floating log one morning and fell victim to an Indian trapper's muskrat trap. She waited until the trapper came, then driving dam. The marked hen selected as her nest site a spot on the southeast side of a small, low island about a third of the way up the flowage. It was dry and warm where the spring sun had melted away the snow and well shielded from view by a small spruce and dense growth of sedge and leatherleaf.

She scraped out a small hollow eight or nine inches in diameter and lined it sparsely with dry sedges. The next day, she laid her first egg. The following day was cold and blustery, and she didn't visit the nest. On the third day and each day thereafter, however, she deposited an egg each morning. As the clutch began to build up, she pulled down from her breast; by the time she completed laying her 10 eggs, the nest was completely lined and even had a small roll of down around the rim.

No. 827-00239 began incubating about 24 hours after laying her tenth egg. Whenever she left the nest from this time on, she used the down roll to conceal and insulate the eggs. During the laying period, the drake had established a territory and waiting site in the vicinity of an old muskrat house about 200 yards up the flowage. Here he loafed, fed, preened, and waited for the female to join him. He drove off another pair of blacks that visited the marsh one day on a house hunting mission of their own; they settled finally on a nearby beaver pond.

All went well until the third week in May when No. 827-00239 was about 10 days into her 28-day incubation period. Then the rains came—and came —and came. The holes in the old dam were too small, and some became plugged with debris. The marshy flowage became a lake, and the 10 eggs drifted about among the leatherleaf, chilled and dead.

" . . . antics calculated to win a black duck's heart."

left alone. The next day she joined a small migrating flock with several unpaired males and was soon being courted by two young drakes that vied valiantly for her favors. She accepted the more dominant of the two within a few days, and the new pair resumed home hunting.

Before the first week of May had passed, they had settled alone on a small marshy flowage, not far from Brownville, that was left by an abandoned log-

For several days, the pair wandered rather aimlessly. Then the attentions of the faithful drake stimulated a reaction from the duck. Hormones flowed again, and a new nest site was established on the same beaver pond where the pair had settled that 00239's mate had chased from their first nesting area. A major difference prevailed, however, in that the other hen's clutch was almost ready to hatch, and her mate had already departed for some un-

"After a short cruise . . ."

known moulting site to lose and later regrow in utmost secrecy his powerful flight feathers.

THE STRAIN of the first nest had told on No. 827-00239, and her second nest among the weeds on the ancient beaver lodge contained but eight eggs when she began incubating on the twenty-first day of June. This late in the season, annual vegetation growth was well advanced and made the nest beneath it almost invisible. Furthermore, the old beaver house was surrounded by water at least two feet deep and was situated a good 100 feet from shore. No egg-eating raccoon, fox, or skunk discovered it. Crows seldom visited the flowage, and the mink that approached one day was distracted by a frog and went on without detecting the nest.

At long last, the nest hatched—at least seven eggs hatched on July 18. One egg was infertile. Perhaps the drake's time was past also, or, possibly, he carried a high level of DDT in his system. In any event, he'd left for a moulting area long before the seven young hatched. No. 827-00239 was faced with rearing her first brood on her own.

The weather was warm in late July, and the ducklings hatched, dried off, and left the nest within hours after the first egg pipped. After a short cruise across the upper end of the pond, the hen brooded the young the first night under a dense stand of ferns on the shore. At first light the next morning, the brood followed its mother through the flooded willow bushes and soon learned that gnats, insect larvae, water beetles, and almost anything small enough that moved was good to eat. There was an abundance of insect life in the beaver pond, and the ducklings grew rapidly, feeding long hours on the high protein diet in midsummer. By the time they were two weeks old, the patches of yellow down on the young were already starting to darken; by four weeks, the birds gave the impression of being much darker brown all over.

There was no problem or conflict with the other brood, which had hatched in early June and was already learning to fly at seven to nine weeks. The older brood left for good, on the wing, about mid-August, and No. 827-00239 followed with her youngsters by going over the dam and down the outlet. This proved to be a hazardous trip, however, and one duckling was taken by a large snapping turtle lying in the brook almost in sight of the beaver dam. Another fell to a fox as they crossed another beaver dam further down, leaving only five of the seven when they reached the main river. In the meantime, the hen progressed into her eclipse moult and fed little.

By the time the little family arrived on the river, the old bird was essentially flightless, and the young were testing their new feathers and making short flights. It was early September, and in a matter of days, the young were fully on the wing and joined another flock of young blacks. This new and larger group soon moved to the ripening rice beds at Costigan, leaving No. 827-00239 to shift for herself along the weedy, marshy shores. She travelled slowly down the Piscataquis, and it was nearly two weeks after her brood had left that she reached the main stem of the Penobscot River. Her flight feathers were nearly grown by now, and she could make short flights.

It was at this point that she moved up river and succumbed to the lure of cracked corn and entered the banding trap where we picked up her story in the beginning.

EPILOGUE: As far as we know, No. 827-00239 is still alive and trading up and down the flyway each spring and fall. Perhaps some day in the future we'll learn more of her whereabouts from a report on the band she carries. The chances become slimmer each passing year, however, and not everyone reports a banded black duck. Foxes, owls, mink, and other predators take their toll. Lead poisoning, disease, and oil pollution are ever present hazards.

Despite the many circumstances that reduce average black duck life to something less than two years, we have records of 10-year-old birds. We know of one ancient female that lived at least 20 years from the time she was banded on the Penobscot until brought down by a hunter's gun on the marshes of Chesapeake Bay in Maryland. So — though the odds are against us, possibly we will write a sequel to this present tale. ∎

Blue-Wings

**THE LIFE HISTORY OF THE INTERESTING
AND FAMILIAR BLUE-WINGED TEAL**

By Gary G. Donovan

SPRINGING ... TWISTING .... TURNING! — the swift, erratic maneuvers of the blue-winged teal are quickly recognizable by waterfowl hunters. The teal's willingness to decoy readily offers tempting shots, but even the most capable gunner ejects spent shells in disgust after underestimating the uncanny agility of this pint-sized dabbler. In Maine, the blue-wing has long been established as an early season favorite of sportsmen.

Taxonomically, the blue-winged teal belongs to the same genus as the familiar black duck, mallard, and green-winged teal. Ducks of this group characteristically confine their activities to wetlands lacking extensive areas of open water, such as streams, ponds, and shallow marshes. They are capable of diving but usually feed at the surface or by tipping rather than submerging. When frightened, these dabbling fowl spring vertically into the air instead of using the running take-off pattern so typical of the divers.

The blue-winged teal is Maine's only small marsh duck with large chalky-blue patches on the fore edge of the wing. In his breeding plumage, the male is dressed out in a pinkish-cinnamon color marked with round dark spots, a slate-grey head, and a conspicuous white crescent in front of the eye. The less colorful female is described as a mottled, buffy-brown duck, with a white breast. About June, the adult male begins to moult the colorful breeding plumage, and by August, his "eclipse" phase closely resembles the coloration of the female. He retains this drab attire until late fall and long after all the blue-wings have headed for warmer climates.

FOR THE most part, the blue-winged teal doesn't even consider returning to points north until spring is well advanced and warm weather is the rule rather than the exception. Some arrive home "paired," but most generally return in small flocks or courting parties where as many as four males may try to win

Male blue-winged teal.

© Leonard Lee Rue III

DUCK IDENTIFICATION GUIDE FOR HUNTERS

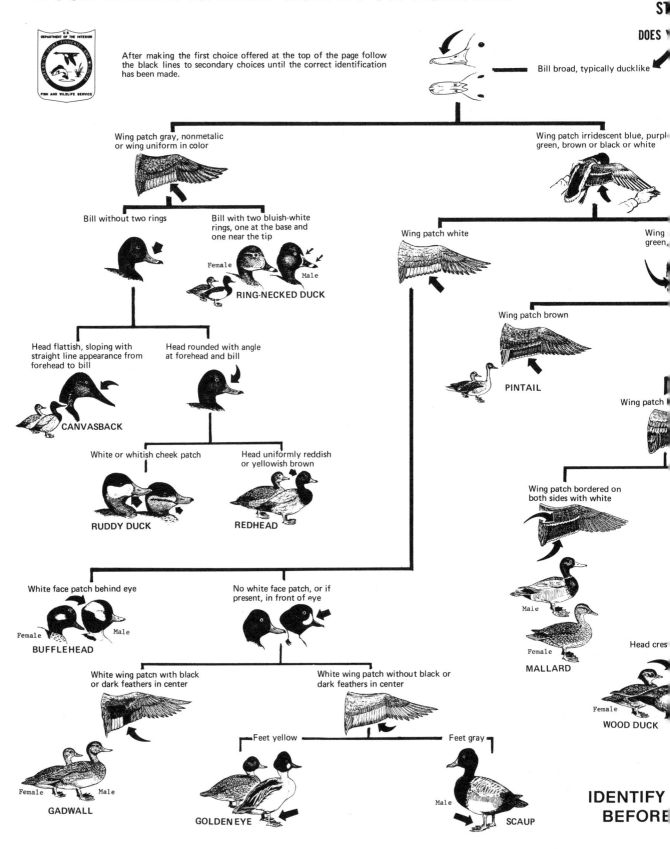

After making the first choice offered at the top of the page follow the black lines to secondary choices until the correct identification has been made.

Bill broad, typically ducklike

ST

DOES

Wing patch gray, nonmetalic or wing uniform in color

Wing patch irridescent blue, purpl green, brown or black or white

Bill without two rings

Bill with two bluish-white rings, one at the base and one near the tip

Female   Male
RING-NECKED DUCK

Wing patch white

Wing
green,

Wing patch brown

PINTAIL

Wing patch

Head flattish, sloping with straight line appearance from forehead to bill

Head rounded with angle at forehead and bill

CANVASBACK

Wing patch bordered on both sides with white

White or whitish cheek patch

Head uniformly reddish or yellowish brown

RUDDY DUCK

REDHEAD

Male

Female
MALLARD

Head cres

Female
WOOD DUCK

White face patch behind eye

No white face patch, or if present, in front of eye

Female   Male
BUFFLEHEAD

White wing patch with black or dark feathers in center

White wing patch without black or dark feathers in center

Feet yellow

Feet gray

Female   Male
GADWALL

GOLDENEYE

Male

SCAUP

IDENTIFY
BEFORE

32

ERE

CK HAVE

Bill slender, pointed, and toothed

Feet yellow or yellowish-gray

Feet pink or reddish

Female | Male

HOODED MERGANSER

COMMON MERGANSER

RED-BREASTED MERGANSER

lic blue, purple, black

ng patch blue, purple, green or black

ole

Wing patch without white border or white only at feather tips

Wing patch green or black

Blue patch on shoulder of wing

Patch on shoulder of wing not blue

low

Head not crested, feet orange-red or coral red

Bill very large and broad, feet orange or coral-red

Bill normal, feet yellow

Shoulder of wing gray or brownish

Shoulder of wing with white patch

le

BLACK DUCK

SHOVELER

BLUE-WINGED TEAL

Female | Male
GREEN-WINGED TEAL

Female | Male
AMERICAN WIDGEON

This pictorial aid is designed to assist in recognizing ducks in the hand after they have been bagged.

The shape of the bill, wing markings, color of feet or head crest are some of the typical characteristics used to identify ducks in the hand. This is quite different from identification of ducks in flight or sitting on water. When flying or on water other identifying features are used such as silhouettes, mannerisms of flight, wing beat, speed of flight or color patterns on body and wings. Every effort should be made to learn to recognize ducks before they are shot. By doing this the hunter is able to take much greater advantage of his sport.

Although occasionally seen inland, sea ducks are not included in this key. They are most frequently found in open salt water areas.

U. S. DEPARTMENT OF THE INTERIOR
March 3, 1849

Cinnamon teal is similar to blue-wing teal except that male cinnamon teal is reddish on head and underparts. The female is virtually identical to the female blue-wing teal.

Female American widgeon has brown breast and flank. Female green-wing teal has gray speckled breast and flank.

DUCKS
OTING

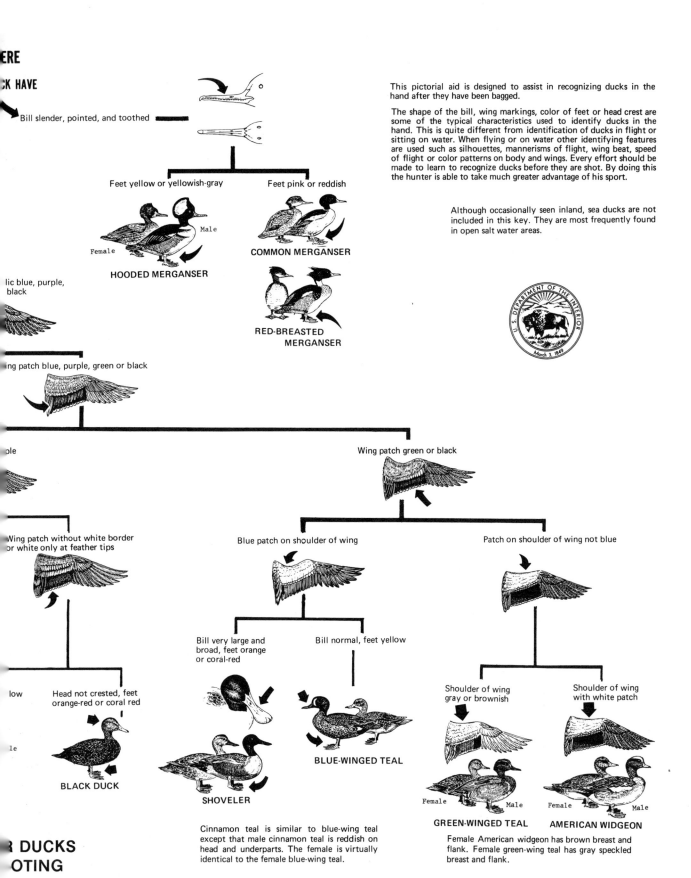

33

a single female. Eventually, one male wins out over the others after a considerable display of courtship antics. Since courtship largely occurs during the migration period, it is not surprising that many of their displays are carried out in the air while coursing and dipping over waterways in a manner similar to that of the black duck. The most intriguing courtship behavior occurs on the water when the pair repeatedly bow to one another over extended periods of time. All the while, the male is on the alert to chase intruders from the female.

Upon completion of the courtship activities, the pair make a critical inspection of many areas before establishing their "territory." Any other members of the same species will be driven from their chosen site. As the time approaches for egg laying, the female selects a nest site and starts construction. Blue-wings have been observed nesting in scattered areas throughout the eastern and northern portions of Maine. Generally, they prefer to nest in the grassy edge of a field or meadow adjacent to water.

The nest is usually a slight depression scraped out of the soil and lined in a bowl fashion with loosely matted, dead grasses and other available dried vegetation. As a final touch, the rim and sides of the nest are insulated with fluffy down pulled from the hen's breast. This is intermixed with the grasses to form a concealing blanket that is easily pulled over the eggs whenever the hen leaves.

In many cases, egg laying actually begins before the nest is completed. Under normal situations, one egg is deposited each morning until a clutch of 6 to 15 eggs is completed. Each day, after the egg has been laid, the hen rejoins the waiting drake on the established "territory." On returning to the nest the following day, the male often accompanies his mate in flight over the site. The cautious female will generally alight at some distance and sneak to her nest. The drake attempts to camouflage her sudden departure by continuing a well sustained flight to his "territory."

Incubation of the eggs begins within 24 hours after the final egg has been deposited and lasts 21 to 23 days. Throughout incubation, the female leaves her clutch only for short periods of feeding and exercise. By the third day, the male begins losing interest in the affair after being given the "cold shoulder" and flies off to join other males in similar circumstances. As a rule, these males do not associate with the females or their broods until the young have been reared and are flying.

During this period of separation, the males seek densely vegetated areas to exchange their colorful breeding plumage for the less attractive female-like "eclipse."

A few hours after the first egg is "pipped," or about to be hatched, all of the ducklings will have struggled from their confining shells. In another four hours, they become dry, and they are strong enough to be cautiously led by their devoted mother to the nearest water. Here they are rapidly taught to feed on soft insects, worms, and other tender animal foods. As they grow and mature, the diet changes to one consisting mainly of sedges, pond weeds, grasses, and smartweeds. During the following weeks of development, the hen blue-wing remains close at hand teaching where to find the choicest foods and diligently guarding the safety of her young by distracting intruders with daring and virtual disregard for her own welfare.

Young blue-wings develop exceptionally fast, as they are late to hatch and early to migrate. In fact, by the sixth week, the ducklings are so active and advanced that they are learning to fly! Once the young are flying, the close family ties are broken, and the young blue-wings disperse from their first summer's home to concentrate with others of the same species in preparation for the fall migration.

During the final days of summer and as migration nears, the blue-winged teal are often seen congregating in flocks of increasing numbers. They begin to concentrate in Merrymeeting Bay during the last weeks of August and reach a peak concentration of many thousands toward the third week of September. By the onset of the hunting season, however, the great majority have journeyed far from the sights of the Maine waterfowler's gun.

These remarkable teal are the first of the dabbling ducks to migrate south in the fall. They commonly winter in the Gulf states, Mexico, Central America, and the West Indies, and frequently travel as far south as Brazil, Equador, Peru, and Chile in search of suitable shallow lagoons. An immature male teal banded September 10, 1969, at Merrymeeting Bay was recovered in the West Indies on October 3, 1969, only 24 days later! Three other similar recoveries were also made the following day.

THE EARLY FALL migration pattern of the teal has exposed this game bird to relatively light hunting pressure throughout the United States. Biologically, the blue-wings apparently can support additional shooting without jeopardizing their populations. A special early teal season will afford Maine's sportsmen the opportunity to enjoy additional days of duck hunting and to harvest this high quality waterfowl species. ∎

HONKERS!

By Howard E. Spencer, Jr.

L ISTEN!—HONKERS! What is the magic of this word and of the bird? Be he school boy trudging home, a mechanic bent over a balky engine, a hardened hunter or a scholar immersed in thought, who doesn't pause to look up, to count, to wonder? Some of us are supposed to be scientists—to know what makes the Canada goose tick, how to make him come to our marshes in greater numbers, how to solve his problems of lead poisoning, how to overcome his over-fondness for sprouting grain, how to preserve him for our grandchildren to enjoy. May God grant us the wisdom to carry out these tasks and forbid our ever making him a pauper or just another target flying by.

I like the words of Francis Kortright, who says of this great game bird, "Sagacity, wariness, strength and fidelity are characteristics of the Canada goose which, collectively are possessed in the same degree by no other bird. The Canada in many respects may serve as a model for man." And Arthur Bent, who adds that, "It is so wary, so sagacious, and so difficult to outwit that its pursuit has always fascinated the keen sportsman and taxed his skill and ingenuity more than any other game bird."

In the same breath, the bird lover would be well advised not to condemn the sportsman-hunters and to recall that it has been and will be their appreciation for this great bird that generates the needed research, management, and purchase of habitat and that foots the bill for most of it.

Unlike many species, when we strip away the romanticism and emotionalism associated with Honkers and try to look upon them objectively, they still command great respect. The ornithologist knows the common Canada goose as *Branta canadensis canadensis*—*Branta* being a corruption of Greek *Brenthos*, the Aristotelian name of an unknown bird, and *Canadensis,* a Latinized form meaning *Canada.* Research by a number of workers has resulted in further subdivision of the Canada geese into several races or sub-species. At least five sub-species of the larger Canadas are commonly recognized (*B.c. occidentalis, moffitti, parvipes, leucopareia* and *interior*). In addition, the giant Canada goose (*B.c. maxima*), weighing upwards from fifteen to nineteen pounds and long believed to be a myth, has been recently re-discovered by Dr. Harold Hanson of the Illinois Natural History Survey.

The opposite end of the size scale is represented by the small Richardson's (*B.c. hutchinsii* Richardson) and Cackling geese (*B.c. minima* Ridgeway), the latter weighing little more than a big mallard or black duck. For the most part, the differences among these races consist of distribution, size, behavior, and minor variations in color shading. All are clearly recognizable as Canada geese.

Maine is probably host to both the *B.c. canadensis* and *B.c. interior* forms although no clear-cut study has been made to determine the local distribution and abundance of the two forms. Though both are essentially of equal size, some data indicate the *canadensis* or North Atlantic population may average a whisker larger. The latter form tends to be somewhat whiter of breast and rump compared to the "grayer" *interior.* From the gunner's standpoint, there is no reason to choose one above the other, but there may well be unknown though important characteristics of each, which have bearing upon their management. Both weigh from seven to thirteen pounds.

THE BREEDING range of the Canadas extends from the Atlantic to the Pacific across sub-arctic and temperate Canada. Important breeding populations also occur in many of the more northern states. Their winter range covers much of the United States, and they often remain as far north as food and water are available. The North Atlantic population tends to be coastal and more marine than the *interior* form. Some of these birds remain along the Nova Scotia coast all winter, and geese observed in Maine during the winter are probably part of this population as are the coastal fall migrants. The flocks which move through the Ft. Kent-St. Agatha region and thence southward across central and western Maine may quite likely be the *interior* form coming out of Québec north of the St. Lawrence. Much more study is needed to appraise accurately the fall goose movements through and over the Pine Tree State.

Both groups come together and winter on the traditional grounds from Maryland to North Carolina. Though they utilize the same general winter range, there is a strong social unity within the flocks, and as far as is known, there occurs little if any intermingling between the two races. In addition to the strong tendency to remain amongst their own kind, there is also a close family bond, with parents and goslings usually remaining together until they return to the breeding ground in the spring. This return to the natal areas occurs early—often before the ice has completely left. Geese have been observed crossing from coastal Merrymeeting Bay to the St. Lawrence River in March, when three feet of ice topped by three feet of snow covered the lakes of the Moosehead plateau over which they flew.

Once they reach their breeding grounds (frequently the marsh or lake on which they were hatched), the adults immediately go about the business of nesting while the yearlings gather and wander. Though there may be some pairing off by the yearlings, many do not mate until their second year. Even as two year-olds, they rarely actually nest. However, mated sub-adults may play at housekeeping even to the extent of selecting and defending a nest territory. Canada geese are usually three years old before laying their initial clutch and raising their first brood. These great birds are strictly monogamous and traditionally mate for life.

Characteristically, geese are creatures of seclusion during the nesting period. This aversion to neighbors has favored their status and stabilized their annual productivity, by comparison with other species. Rarely if ever do catastrophic losses occur over any appreciable proportion of their breeding grounds. The lack of man's influence upon much of the northern breeding habitat has also thus far been a saving grace. Nesting densities tend to be sparse, and twenty pairs per square mile is a veritable "crowd."

Although commonly ground nesters, Canadas sometimes select abandoned hawk or eagle nests in which to start their brood. Man has been successful in encouraging them to nest in a variety of elevated artificial structures protected from predators and flooding. Frequently, an abandoned muskrat house is chosen as a nest site, and

when the clutch of four to ten eggs has been deposited, it is fiercely defended by both parents against all comers. The gander remains nearby throughout the twenty-eight-day incubation period, and when the goslings hatch, he continues to help rear and protect the brood.

Soon after the goslings arrive, the parents moult their wing pinions and are flightless for about three weeks. The timing is such that the old birds regain the power of flight about the same time the young learn to fly. Once they are on the wing again, it is not long before the fall migrations begin; and though the rear guard may not pass through Maine until mid-December, the first honkers may be expected anytime after mid-September.

Geese are inclined to be more terrestrial than their aquatic cousins, the ducks, and feed to a considerable extent by grazing. Succulent green sprouts of almost any kind are a preferred food. Waste grains of all kinds left by the harvesters are heavily utilized. In Maine, small potatoes left in fields after picking are greedily consumed, and there are records elsewhere of Canadas consuming sugar beet tops left following the harvest. Many natural foods are taken, according to availability. For Maine's coastal flocks, eel grass is a choice and abundant item, while at Merrymeeting Bay, wild rice, bulrushes, and numerous other aquatic plants are avidly consumed. As in ducks, during the first few days and weeks of the goslings' life, insects are a major constituent and an important source of protein in the diet.

HUNTING THE CANADA goose may be considered the epitome of wildfowling. They are pursued in many different ways and under many different conditions, from the sportsman in the cornfields of Maryland to the Eskimo on the tundra in Labrador. In Maine, they are hunted largely over decoys in stubble fields or on the shallow, rice-filled waters of Merrymeeting Bay. Unfortunately for Maine gunners, many geese apparently overfly Maine non-stop during the fall migration, and hunting opportunities are decidedly limited. If they do stop, it is often but very briefly, and they may be many miles south twenty-four hours later. Exceptions to this pattern occur in the Ft. Kent-St. Agatha area, at Merrymeeting Bay, and in the Taunton-Egypt-Hog Bay area of Hancock County. Sizeable flocks utilize these areas quite consistently each fall.

Goose hunting in Maine is a sport for the trophy seeker, and considerable effort and skill are usually required to bag a 'honker.' The Canada goose harvest in Maine has ranged from 500 to 1600-plus in recent years. The estimated retrieved kill in 1965 was the largest on record and attests to excellent status of the Atlantic Flyway population. Presently, Canada geese wintering in the Atlantic Flyway after the hunting season are numbering in excess of half a million. By comparison, this is about double the number of black ducks.

Let's take a look at the management and preservation operations that affect geese. Fortunately, in the present day and age, we have an organization, international in scope, called the Atlantic Waterfowl Council which is concerned with (among other things) the management of geese. Composed of game scientists and administrators from eastern Canada and United States, this organization has prepared management and research plans which aim at still further improving the status of Canada geese and have set a goal of one million geese in the post hunting season Atlantic Flyway population. Additional management objectives include improved population and harvest distribution, plus improvement of the recreational quality of goose hunting.

To attain these stated goals, well co-ordinated and integrated research and management programs will be necessary. Many such programs are already well underway. Within the last two years, successful techniques for banding geese on the eastern Canadian breeding grounds have been developed and a $70,000 co-operative annual banding program for geese and ducks initiated. Information from recoveries of birds banded on their natal areas will enable determination of where and in what degree birds from particular breeding areas are being harvested. Similarly, wintering ground bandings will enable association of breeding grounds with specific wintering flocks, as well as measuring where the harvest of the wintering flocks occurs.

The importance of such data can be appreciated when one considers that the same flock of geese may be hunted near its Canadian breeding ground in September, on the oat fields of Ft. Kent in October or November, and in corn fields from Maryland to North Carolina during December and January. Management needs to know which flocks are being harvested to the limits of their ability to produce, and which ones might withstand additional shooting without serious consequences. Concurrently with banding programs, harvest regulations have been altered and, the daily bag limit reduced by one-third from three to two per day. This is aimed at reducing the overall kill and returning more birds to the breeding ground to build up that million bird population.

It is noteworthy that this restriction in harvest has come into being at a time when the population is flourishing rather than when things are at a low ebb. Careful harvest studies are being used to evaluate the effect of hunting regulations as a management tool. In addition to banding and harvest regulations, population distribution is being studied and habitat surveyed to determine how best to disperse and more equitably distribute both harvest and populations. Artificial propagation and transplanting are methods being used to establish decoy flocks that attract migrants to hitherto unused areas. New local breeding flocks can also be successfully initiated in suitable habitat. Once these flocks are firmly established, their production adds to the basic population and harvest.

From the management viewpoint, better distribution of populations not only improves harvest opportunities but may also serve to alleviate agricultural depredation problems and reduce the incidence of disease or lead poisoning. The latter has been an increasing mortality factor in recent years. Geese feeding in areas which have been heavily shot over, ingest the shot; significant numbers die as a result.

The usual clutch of Canada geese is from four to ten eggs. The nest is guarded by both parents.

MAINE IS just beginning to involve itself with goose management. The Moosehorn National Wildlife Refuge at Calais has been successful, after many years work, in starting a local breeding flock. The Game Division of the Department of Inland Fisheries and Game has transplanted nearly three hundred honkers from a resident flock in New York during the last two years. It is hoped that these birds will act as decoy flocks, encouraging migrants to remain longer in Maine during the fall and eventually, with careful nurturing, to establish one or two local breeding flocks. During September 1966, a team comprised of representatives from the Department of Inland Fisheries and Game, the Wildlife Management Institute, the U. S. Bureau of Sport Fisheries and Wildlife, and the U. S. Soil Conservation Service investigated many of Maine's most favorable potential goose management areas. As a result, plans are currently being formulated to institute various management measures on several. Meanwhile, the banding programs, harvest studies, and population censuses which provide our base of knowledge from which to work will continue.

Last but not least, it is natural in this day of the ever expanding megalopolis to ask what the future holds for the Canada goose. As the reader will have already detected, there is good reason for optimism. The major breeding grounds are large and remote and will probably be little affected by man's activities in the foreseeable future. The species responds well to skillful management and there is a host of dedicated sportsmen, administrators, and game scientists devoting money and energy to seeing that this management is provided. In short, there seems to be no reason why we should not be hearing the melodious cries of this greatest of game birds for generations to come, as he moves up and down the North American continent following the traditional paths of his ancestors.

Photo by Arthur Rogers

Don't Slight the Snipe!

By Howard E. Spencer, Jr.

"Now HOLD the bag in this little run with the flashlight just above it. We'll drive the snipes into it, then we'll have a feed to remember. Don't give up though, 'cause sometimes it takes quite a while to get 'em coming." With those words we left him at the edge of the frog pond in back of old Applesauce Hill. It was 'most midnight 'fore he got back and he was *some* mad fer a while. (Didn't get no snipes neither.)

I wonder how many of our readers have shivered in the spooky dark of an unknown swamp waiting for the elusive snipe? Certainly this little game bird has furnished lots of sport for lots of people over the years. They'll lie close for a dog and fly like a corkscrew. Here today, gone tomorrow, they test the scatter-gunner sorely.

For the benefit of unsuccessful night hunters, snipe are small, brownish, long-billed birds weighing less than two ounces and only twelve inches long. They're known as jacksnipe to the hunter, Wilson's snipe in the bird books, and *Capella gallinago delicata* (Ord) to the scientist. Groups up to twenty or forty are called "wisps" and larger congregations are termed flocks.

Distributed over much of North America and northern South America, snipe breed from northwestern Alaska, northern Mackenzie, central Keewatin, northern Ungava, and Newfoundland south to northern California, southern Colorado, northern Iowa, northern Illinois, Pennsylvania, and New Jersey. They winter in California, New Mexico, Arkansas, North Carolina, through central America and the West Indies to Columbia and southern Brazil. There are also records of jacksnipe in the Hawaiian Islands, Bermuda, and Great Britain.

The snipe's food includes fly larvae, aquatic beetles, crustaceans, earthworms, snails, and small fish. Some vegetable items, such as smartweed, bulrush, sedges, and wild millet, are also eaten.

The spring courtship flights of snipe have intrigued observers for decades, and there has been much debate over the origin of the "winnowing" or "drumming" sounds made during brief aerial power dives. Since "winnowing" usually occurs at considerable height it is difficult to detect whether the unusual sounds are produced by vibrating wings or some other source. It is now generally agreed that two special feathers on either side of the tail give rise to the drumming.

A monogamous species, snipe nest largely in low, wet meadows, bogs, or swamps. Usually, grassy cover is selected for the actual nest site, and three or four pointed eggs are deposited. Egg color is grayish olive, spotted and streaked with chestnut, burnt umber, and black. Incubation lasts nineteen - twenty days. The young are downy and leave the nest soon after hatching. Probably, it is unusual if more than two young survive to flying age.

Snipe hunting with dog and gun is apt to be unreliable due to the migratory nature of the species. Furthermore it frequently requires considerable physical stamina and agility to negotiate the soft boggy areas that snipe frequent. For hunters who enjoy something different, however, the little jacksnipe may on occasions provide much sport and offer challenging targets. Hunting is regulated by federal law which prescribed a 1966 season from September 26 through November 14 and a daily bag limit of eight (possession limit sixteen). Shooting hours are sunrise to sunset. Harvest studies indicated that about 800 hunters shot 1,500 snipe during Maine's 1963 season. ■

References

Bent, Arthur C., 1927. *Life Histories of N. A. Shorebirds.* Dover Publications, Inc., N. Y. 14, N. Y.

Job, Herbert K., 1917. *Birds of America.* Garden City Pub. Co., Garden City, N. Y.

Pollard, Hugh B. C. 1936. *Game Birds and Game Bird Shooting.* Houghton Mifflin Co., N. Y. 36, N. Y.

Roberts, Thomas S. 1936. *The Birds of Minnesota.* Univ. of Minn. Press, Minneapolis, Minn.

The male
American woodcock
is a real
showoff
when
it's time
for
courtship

Timberdoodlings

By J. William Peppard

WHILE THE American woodcock or timber-doodle is best known for his sporting qualities as a game bird, he is also fascinating during his spectacular courtship period in April and May. His ritual of songs and flights is observed by many people in Maine, especially rural folks, who look forward to these antics as one more positive indicator of spring.

After escaping the northern winter by flying to the lower Mississippi Valley, the woodcock — *Philohela minor* — heads back toward Maine and points north. It's generally believed that the male precedes the female in the spring migration.

Back in Maine, the male chooses his territory or courtship and nesting area, to which he claims ownership by his aerial acrobatics. The territory or singing ground is usually an open area, such as a pasture or field, near a stand of hardwood or mixed growth. Both the male and female use the wooded area for protection during daylight hours and for a feeding and nesting place. The woodcock's diet consists of earthworms (more than 80 per cent) and also grubs, beetles, and larvae.

Once established on his spring territory, the male begins a daily two-act performance which takes about a half-hour, at daybreak and dusk. Conditions such as rainy or snowy days or moonlit evenings may cause variations in this schedule.

The exhibition the male puts on, and his singing, are believed to attract and hold his female and also to defend his area from intruding or jealous males. The female is seldom, if ever, seen on the singing ground during the displays. She is believed to be most often in the protective cover of the nearby woods and nesting area.

The woodcock's appearance on the singing ground is very punctual — apparently governed partly by the intensity of the light. He usually starts in the evening when about two candle power of light remains, and usually stops in the morning at about the same light conditions. Therefore, on a cloudy evening he may start a little early; and on a bright, moonlit night, he sometimes finds it difficult to stop.

The first act to be seen occurs when the male flutters into the singing ground; he may be observed silhouetted against the darkening sky. He lands and selects a spot in the opening, perhaps about four feet square. This point he will use as a stage — a take-off and landing strip — during the next thirty or forty minutes.

Soon, he will sound a series of "peents," which have been described in many ways; they are like the angry buzzing of a bumblebee in one's ear, ending with a very abrupt *b-z-z-t!* This peent is repeated up to fifty times while the male walks around his take-off strip. He faces different directions as he calls; hence, he has been described as throwing his voice or being a ventriloquist. Depending on which direction he is facing, some of the peents will seem quite distant and others very near.

Most people can easily hear and recognize these calls, but others, who may have just a slight hearing

A female woodcock sits on her nest, hidden, from all but the sharpest of eyes, by her protective coloring.

defect, may find the peents next to impossible to hear.

Finally, the calls increase noticeably in tempo, and the bird takes to the air to demonstrate his acrobatics. He is easily spotted against the sky as he rises above the tree line, heading for the top of his domain in ever-decreasing circles, to an altitude approaching three hundred feet. During this upward flight, the bird's flight feathers make a whistling sound.

After arriving at the height of his flight, usually over his singing ground, the woodcock generally makes a few brief turns, perhaps to declare his ownership of space. Then he heads for his landing strip, again flying in ever-decreasing concentric cir-

cles and making a very melodious chirping. When he is forty or fifty feet above the ground, the star of the show simply tucks in his wings and dives for the landing strip. Somehow, he manages to brake his descent just before touching the ground.

This landing completes one flight, and the woodcock immediately starts peenting and preparing for a second performance. The next series of peents, though, is not as long as the first, and the acrobat doesn't stay on the ground as long as he did before his first flight. Each flight lasts about one minute and may be repeated ten to fifteen times during each morning and evening singing period.

To see this performance, one should plan to be at or near an opening or field about one-half hour after sunset. As long as the observer remains quiet, the woodcock apparently doesn't mind having an audience. While the male is in the air, the observer may move, quickly and silently, closer to the take-off point. After several such moves, one may be within fifteen or twenty feet of the bird when he lands. It is important to remain very quiet and close to the ground. If the performer is frightened, he will fly off to another point or remain silent for the rest of the singing period.

Besides the peent, two other calls are given by the woodcock. A chattering or cackling sound is uttered, both on the ground and in the air, when one male intrudes on the singing ground of another. It is usually given as the male defends his territory, and the harsh sound lets the observer know that a challenge has been made. Both males may be seen, but they usually fly too low for good observation during such an altercation. More often than not, the affair is settled quite easily in mid air, and the intruder is driven off to greener pastures!

The other call, heard while the male is on the ground, is a gurgling or swallowing sound which is quite difficult to hear, recognize, or describe. It usually precedes the peenting call, and no particular purpose for it is known.

Just-hatched chicks, too, are protected by being colored much like the grass, leaves, and twigs around them. Even the eggs are quite inconspicuous.

As THE WEATHER WARMS and the days grow longer, the male woodcock increases his courtship activities. Courtship generally starts in April, reaches a peak about the first of May, and tapers off during May and June. Occasional courtship displays occur in the summer, but they are not equal to those in the spring. Also, Maine climate varies enough so that the woodcock performances in southern areas are about two weeks earlier than those in northern or eastern Maine.

Throughout most of the courtship period, the female is seldom seen. On rare occasions, one is observed on the singing grounds, but she does not make regular appearances like the male. She probably spends most of her time near or on the nest, which will ordinarily contain four eggs.

Woodcock nests, usually not very elaborate, are made of leaves or twigs in a slight depression on the ground. Apparently, the wonderful protective coloration of the birds insures against excessive nest losses.

After incubating about twenty days, the eggs hatch throughout May, so that by June, most young woodcock are actively following their mothers. The chicks spend only a few hours on the nest after hatching.

If danger threatens her eggs or young, the female also can turn in a noteworthy job of acting. When the situation warrants, she will feign injury by drooping one wing, calling softly, and making short, awkward flights with her feet dangling, in order to attract the intruder's attention to herself. The young hide under leaves and twigs, their safety dependent on their coloration, which is so perfectly designed that a very sparse cover will camouflage either the young or old birds from their natural enemies. After the danger has been led away, the mother returns to her brood.

Two weeks or so after they hatch, the young can fly nearly as well as the adults. They are increasingly independent and eventually are completely on their own. Following a whole summer of learning the trials and tribulations of being a woodcock, they begin the long migration to the southern United States. Migration usually lasts throughout most of September, October, and early November in Maine.

THE HUNTING SEASON presents more hazards, but the woodcock is a "game" bird, and hunting has become a part of his life history. He has held up remarkably well so far under increased gunning pressure, and he always returns to Maine in ample numbers to provide another successful breeding season. According to the Fish and Game Department's game kill questionnaire, the average annual bag of woodcock in Maine was about 22,000 in recent years. The outstanding courtship behavior of the timberdoodle has helped game biologists to learn more about woodcock populations and activity. In the spring courtship season, a count of the males observed singing may be compared with counts made in previous years, along designated survey routes. The method has not been infallible, but it has helped provide an index of the number of male woodcock returning to Maine each breeding season.

Also, the male's tendency to perform at regular intervals on a given site has given an excellent opportunity to livetrap and band male woodcock. Usually caught with nets of various kinds, they are numbered with a small band of light metal attached to one leg and then released to continue courtship. Instructions on the band request that it be returned to the Bureau of Sport Fisheries and Wildlife, so the movements of the bird after banding may be accurately recorded. Color-markers may also be attached to the birds to provide field identification so their travel activities may be recognized without having to trap them again.

BIG SURPRISES come in small packages." This is surely true of the American woodcock. If you find yourself interested in this story of aerial acrobatics, take some time off on an evening about dusk in late April. Walk down the road, along fields and pastures, and be prepared to be fascinated by the antics of the male woodcock in courtship season. ■

The tiny woodcock chicks spend only a few hours on the nest after hatching. Within a couple of weeks, they are able to fly almost as well as adults.

Woodcock equipped with radio transmit and antenna (arrows) allow biologists to study daily activities. Fluctuations in signals tell whether bird is moving on ground, flying, or rest

ANY WOODCOCK IN YOUR BACK FORTY TONIGHT?

By William B. Krohn
and
Ray B. Owen, Jr.

MORE THAN 1,100,000 woodcock were harvested in the eastern United States and Canada during the 1972 hunting season. In some northern states, including Maine, more "timberdoodles" than inland ducks are shot each fall. Not only is the total number of woodcock taken by hunters impressive, but the popularity of this game bird is growing each year.

The impact of increased hunting on populations of migratory birds can best be measured by studying recoveries of birds banded during the summer months. Recoveries of banded woodcock also allow biologists to investigate migratory patterns (see **Maine Fish and Game,** Spring 1973). During the summer,

woodcock are most easily banded on fields at night. To make banding operations more efficient, and to determine what types of habitats are preferred, we undertook studies to learn what Maine woodcock do between June and November.

Direct observations, and monitoring birds equipped with small radiotransmitters, have shown that woodcock make daily flights from coverts to fields about 30 minutes after sunset. The average distance between coverts and fields is about 1,000 feet, and the flights of individual birds rarely last more than 1 or 2 minutes. Usually, all birds using a field will enter within a 10 to 15 minute period. The birds remain on fields throughout the night and start leaving around 45 minutes before sunrise.

Light intensity apparently stimulates these twilight flights. As the

Both truck-mounted and hand-held antennae are used to monitor radio equipped woodcock. Each radio transmits a different frequency so individual birds can be distinguished.

42

time of sunset becomes progressively earlier between June and November, evening flights into fields become correspondingly earlier. Similarly, as summer progresses and sunrise becomes later, morning flights from fields are delayed. During evenings and mornings when the sky is overcast, birds enter fields earlier and leave later than when the sky is clear.

We have found that forest openings, often old fields, used by woodcock during the summer months are also used by courting males in April and May. In early June, courtship ceases, but the males continue to use clearings. Some females take their broods onto fields at night; and, by the end of June, field use increases greatly as the immature birds attain full flight capabilities. The number of woodcock using nocturnal fields reaches a peak in July, drops during August, increases again by mid-September, and decreases during the middle of October. The mid-October drop in field usage coincides with the start of the fall migration. In early November, few woodcock can be found on fields at night in Maine.

Although we do not know exactly why the number of birds using fields drops in August, we suspect that this decrease is related to the molt. Woodcock replace most of their feathers during August. Possibly, molting birds are less able or less inclined to move to fields during this period.

WOODCOCK USE a variety of openings including abandoned farm fields, young evergreen plantations, blueberry fields, logging roads, power lines, and bogs. We have found a few woodcock at night on the median strip of Interstate Route 95, and in gardens behind houses! Small openings are usually utilized by only 1 or 2 birds per night. Larger openings, 10 or more acres, are occasionally used by 20 to 30 woodcock. On one 15-acre field, between 45 and 50 individual woodcock were flushed during one visit!

Within a field, birds rest in small pockets of low cover surrounded by tall and dense vegetation. Fields with broken cover, such as abandoned farm fields, usually have many protective pockets and are thus preferred over openings with a uniform cover, like hay or cultivated fields. Preferred fields are generally surrounded by extensive stands of young, second-growth, hardwoods including alder, aspen, and birch. Such woody cover provides the female woodcock with areas to nest and raise their young.

Monitoring the daily activities of birds carrying miniature radios has shown that young and adult woodcock behave differently. Immatures tend to be found in the central portions of large fields, while adults frequent small openings and the edges of large fields. Adult woodcock walked to fields more frequently than young birds, which usually flew. In addition, immatures moved greater distances than adults between diurnal coverts and nocturnal fields. Observations during thunderstorms, heavy rains, and high winds indicated that birds make flights regardless of weather conditions.

IT HAS LONG been thought that woodcock go to fields at night to feed. Indeed, woodcock wintering in Louisiana do feed shortly after landing in the evening on large agricultural fields. However, in Maine we found that little, if any, feeding was done on old fields. Examination of stomach contents from Maine woodcock collected before and after landing showed that birds feed heavily on earthworms and insect larvae before flying to openings. Stomachs from birds captured at all hours of the night were generally empty. Also, continuous tracking of radio-equipped woodcock for 24-hour periods revealed that birds rarely moved on fields but were active in coverts immediately prior to the evening flight, just after the morning flight, and sporadically throughout the day. To the best of our knowledge, woodcock in Maine go to fields during summer and fall nights to roost. Possibly, woodcock roosting on fields at night can more readily escape predators than can birds in wooded covers.

The preceding findings have done much to improve the efficiency of banding crews. For example, knowing the time of evening flights as related to sunset and cloud cover helps banders start work promptly. Understanding seasonal fluctuations in the number of birds on roosting fields permits us to predict the most, and least, productive months for banding. In years of normal spring weather and nesting, for instance, banding in August is less productive than in July since usage of fields drops 30 to 40 percent between these months. Knowing that adults tend to use edges of fields may enable banders to increase their catch of older birds. And finally, the lack of feeding in fields suggests that baiting woodcock into openings would be an unlikely approach to capture more birds.

AS OUR STUDIES have progressed, we have become increasingly convinced that heavily used roosting fields are excellent indicators of high local woodcock populations. It is for this reason that we ask: Are there any woodcock in your back forty tonight? If not, then you might consider managing abandoned farmlands and forested areas for woodcock. Our experience has shown that old fields are used not only by woodcock, but that grouse, deer, and other wildlife abound there. As old farms grow from brushy fields into mature forests, game populations decline. For details on how to create and maintain openings, we suggest reading the article on how to "Improve Your Land for Woodcock" by J. William Peppard (**Maine Fish and Game,** Spring 1972). ∎

PART III: NON-GAME BIRDS

This section deals with birds native to Maine for which there is no hunting season—with one exception. Maine does have a hunting season on crows, and these birds are technically considered migratory birds by the federal government. But they are included in this section of this book because the Maine Fish and Wildlife Department does not consider them a game bird in the sense that they are closely managed as a species. It is possible, further, that the dove might be classified as a game species in Maine at some future date (many states have regulated hunting seasons for dove now), but it is a non-game bird at the present time.

TABLE OF CONTENTS

What are the causes of the bald eagle's decline in numbers? What is being done to preserve

OUR NATIONAL EMBLEM

By Edward A. Sherman

IN JUNE 1782, by Act of Congress, the bald eagle became our national symbol. Between 1782 and the year 1940, the bald eagle became a victim of "progress," ominous changes taking place in the status of this majestic bird whose likeness is probably better known all over the world than that of any other living creature. In 1940, Congress, recognizing the questionable future of the bald eagle, announced: "Whereas by the Act of Congress and by tradition and custom during the life of this Nation, the bald eagle is no longer a mere bird of biological interest, but a symbol of the American ideals of freedom."

Behind this adoption of the bald eagle as the emblem of our Republic, is the record of nearly fifty centuries of respect for various members of the eagle family which numbers more than fifty species. Held high as a standard, the eagle was followed by

Roman legions as they conquered the world. Later, the rulers of the Holy Roman Empire, Russia, Austria, Germany, and Poland adopted the eagle as a symbol of sovereignty.

There are two eagles native to our North American continent—the golden eagle, national emblem of Mexico, and the bald eagle. Eagles range over all of Canada and at some time during the year can be seen in all of the forty-nine continental states. During the winter, eagles are found principally in Florida, the middle Atlantic states, the middle west, and the northwest. With the exception of Florida eagles, which are inclined to be resident, these birds are apt to move northward after nesting.

The concern for the future of the bald eagle may date back to 1943 when C. L. Broley, who banded young eagles along Florida's Gulf Coast, noticed a decline in the production of young eagles. In 1947, he found that forty per cent of the nests failed to produce young birds. The decrease persisted, and by 1950 he found only twenty-four young in the 125-mile stretch between Tampa and Fort Myers where he had previously banded 150 eaglets a year; in 1955, there were eight; and in 1958, only one young eaglet was to be found. Except in the Everglades National Park where the bulldozer and other instruments of man have not destroyed the wetlands that favor these fish-eating birds, the picture in Florida was observed to be the same.

Observers such as the Hawk Mountain (Pennsylvania) Sanctuary Association also pointed out the decline in immature birds. Between 1935 and 1940, 38 out of every 100 bald eagles passing over their area were immatures. Between 1953 and 1958, the number of young birds had dropped to 21 out of 100; and in 1963 and 1964, only an average of 10 immatures were seen out of every 100 eagles. In a wildlife population, such a steady decline in numbers of young can spell ultimate extinction.

This decline in active nests and in numbers of immature birds noticed by many observers, led the

National Audubon Society in 1960 to begin an investigation of the bald eagle. This was the first real systematic effort to learn the size and distribution of the bald eagle population in the United States. This study led right to the heart of the matter—low productivity. The finding of pesticide residue in both dead birds and eggs brought into focus an ominous picture of the future of our national bird.

At the same time the National Audubon Society undertook its nationwide survey of numbers, distribution, and nesting success of the bald eagle, the Bureau of Sport Fisheries and Wildlife began a study of the effects of environment and pollution on eagles. The effect of DDT and other chlorinated hydrocarbons on eagles was investigated. So far, such studies reveal that although the exposure of eagles to DDT and other hydrocarbons is nationwide, most eagles in the United States die of causes other than pesticide poisoning. And the important question of sublethal effects of pesticides on eagle behavior, particularly their ability to reproduce, has not yet been entirely answered.

There are several good reasons for the official listing of the bald eagle as a "rare and endangered species." The first and most important is the disturbance by humans of eagle habitat. It is interesting to note that while Charles Broley was noting the rapid decline in eagle numbers in West Florida, the human population explosion figure was reaching 300 per cent. The destruction of wetlands because of the popularity of waterfront lots hurt the bald eagle, and the removal of large trees needed for nesting discouraged reproduction. Shrinking food sources for both parent and young compounded the loss of opportunity for nesting.

WHAT IS the story in Maine? In 1966, Charles M. Brookfield, who made observations in Maine for the Audubon study, found no nesting success west of the Penobscot. Eight eaglets was the total for the State of Maine—all but one of these in Washington County.

Steps have been taken to halt the decline of eagle numbers. Federal laws in the United States protect both the bald eagle and his cousin the golden eagle. These laws are enforced by the Bureau of Sport Fisheries and Wildlife and by state fish and game departments. Meanwhile, the Bureau is continuing its studies of the effect of pesticides on bald eagles.

The National Audubon Society is conducting intensive investigation of bald eagle distribution, status, breeding biology, and other factors. Florida Audubon has set up a Bald Eagle Sanctuary system. Florida landowners on whose property eagles are nesting have agreed to protect eagles and their nests and to refrain from cutting nesting trees.

Personnel of the U. S. Forest Service map the locations of active nests, restrict development within a certain distance of these nests, and prohibit all activity in the vicinity at certain times of the year.

Because the eagle is a creature of space and solitude and needs wilderness conditions, last year Secretary of the Interior Udall announced the establishment of an isolation zone around all active nests on national wildlife refuges.

Records of the Bureau of Sport Fisheries and Wildlife reveal the result of setting aside lands possessing wilderness characteristics and managing them for the benefit of wildlife. In 1962, eighteen eagles were counted within the vicinity of the Moosehorn National Wildlife Refuge in Washington County, Maine, and two nests there produced three young. In 1963, fourteen adults and one immature bird were seen on Moosehorn. In 1964, twenty-two eagles were counted on the refuge. In both 1963 and 1964, hungry eagles raided the captive Canada goose flock at Moosehorn in mid-winter, killing several geese. In 1965, two nests were active there, and in 1966, a nest on the refuge's Magurrewock Marsh raised two young eaglets. Also, this last year at the top of an ancient seventy-foot pine on one of the Birch Islands, part of the refuge in Cobscook Bay, there was an active nest. Only fifteen miles north of Moosehorn, near Pocomoonshine Lake, a nest was observed which produced two eaglets. Eight of our national wildlife refuges in the southeastern United States have bald eagles nesting on them.

Circumstantial evidence continues to point to certain chemical pesticides as serious hazards to wildlife. Because of this, the Department of the Interior, in 1964, ceased using compounds such as DDT, chlordane, dieldrin, and endrin, which are known to concentrate in living organisms.

A campaign of public education is needed to make sure that no bald eagles die of "lead poisoning," for altogether too many of these birds are shot either deliberately or carelessly, since the head and tail of a bald eagle do not become white until the bird is four or five years old. Such a campaign as the one which made people aware of the need for saving the whooping crane from extinction would prevent bald eagles being killed by thoughtless gunners. ■

THE BALD EAGLE IN MAINE

Photo by Frederick Kent Truslow

A STRUGGLE FOR SURVIVAL

BENJAMIN Franklin, who wanted the wild turkey to be our national symbol, probably wouldn't agree that the bald eagle could effectively fill such a role. However, through our history this majestic bird has represented strength and dignity, and now it is serving as a warning of our own possible self-destruction.

Maine is the bald eagle's last frontier in the northeast, and even here it is struggling against insurmountable odds. In 1972, the U.S. Fish and Wildlife Service and the Maine Audubon Society began taking a closer look at Maine's eagle population. Our purposes are: (1) to find and record eagle nest locations, (2) to learn more about the areas in which they are found, and (3) to determine the status of the eagle population.

The first step, before we actually began observing eagles in the field, was to find out all the possible nest locations. Much of our information came from the National Audubon Society which has surveyed the eagle nests annually for the past 10 years. We also checked information recorded in the Natural Areas Inventory, a study done under the auspices of the Natural Resources Council. People from all over Maine, responding to articles in their local newspapers, provided us with more information on possible eagle nest locations. With this data, we were ready to travel throughout the state, running down all leads in search of eagle nests. Incidentally, we are still not sure that all eagle nests in Maine have been charted. If you know of any, please notify the U.S. Fish and Wildlife Service office, Federal Building, Augusta, Maine 04330.

The use of aircraft has proven to be the most effective method of surveying eagles because the nests are located high in trees and often cannot be seen or reached from the ground. Bill Snow, regional pilot and warden for the Bureau of Sport Fisheries and Wildlife, was instrumental in conducting this survey. His Beaver airplane, outfitted with pontoons, served us well. We chose the middle of April to begin the survey because at this time the leaves have not yet come out on the trees, thus making it easier to observe the nests. At this time, also, most of the female eagles have laid their eggs and are just beginning to brood. There were nearly 100 reported locations to check throughout the state; by the time we completed our spring series of flights, we had verified 65 of them.

The second part of the survey was done in late June. By this time, the eggs have hatched and the young eagles are sitting on the nest, easily seen from the airplane. Although we observed only eight young eagles, we were overjoyed to find at least that many, for we knew the number would be low. Having analyzed the results of both series of flights, we estimate that there are about 50 eagles in Maine. Although this number is up a bit from records of the past 10 years because of a more extensive study, it still cannot compare to the estimated 200 at the turn of the century or the estimated 100 in 1950.

In our survey, we found that although most of the eagles nest along the coast, the ones living inland have a much higher reproductive rate. The 18 eagles living north of Bangor produced five young, compared to the 30 eagles living in coastal areas, which produced only two. The remaining eagle was born in Androscoggin County and disappeared from the nest before it was able to fly. We suspect that it was a victim of predation.

IN ORDER to find ways to protect eagles, we need to have an understanding of their life history. A pair of eagles mate for life, but if one dies, the survivor may take on a new mate. Eagles tend to use the same nests year after year. In some cases, they have more than one nest in an area, repairing and adding to them each spring. The nest is made of branches of limbs, lined with twigs, grass, and pine needles. Nests sometimes measure up to 10 feet in diameter and are found in the tops of tall trees, commonly white pine in Maine. The bald eagle always chooses a nest location near a body of water. Under prime conditions, eagles lay two to three eggs per clutch, but in Maine, one egg is most often the case. Incubation takes about 35 days, and it will be at least 10 weeks before the fledgling leaves the nest. The immature eagles are basically black, and three to four years will pass before they obtain the white head and tail of the adult plummage.

The eagle is essentially a scavenger, and its primary food is fish, picked up along the shore or stolen from

47

an unwary osprey. Other parts of the eagle's diet are small rodents, reptiles and amphibians, and larger animals that are either weak, diseased, or already dead. Eagles do not kill healthy deer or livestock as was once supposed. We now know that an eagle can carry only small animals.

Some factors are especially significant in the death of eagles. Of course, there are many causes of eagle deaths, such as impact injuries (cars), electrocution, diseases, and pesticide intoxication. However, one cause is a distinct standout. A study done in 1965 revealed that 62 per cent of eagle deaths are due to gunshot wounds.

Eagle shooting started very early in the American history. Eagle bones have been found in Indian shell heaps. In 1806, the townsmen of Vinalhaven placed a 20-cent bounty on eagles, as the bird was considered a menace and evil predator. But by 1940, eagle numbers had decreased to such an extent that a federal law was passed to protect them. This law carried a maximum sentence of $500 and/or a six-month jail sentence. Congress recently passed a law which increases the penalty for killing or molesting bald or golden eagles. The fine for a first offense of knowingly or with wanton disregard causing the death of an eagle may now be up to $5,000 and/or a year in jail. And these penalties may be doubled for a second offense.

Although eagles are shot every year, to the best of our knowledge no one in Maine has ever been prosecuted for killing one. Many times, the shooting of an eagle is a case of mistaken identity but not always. Gun owners are urged to learn how to identify eagles.

Besides the direct causes of death, there are two indirect influences which reduce the eagle population: coastal development and pollution of our waterways. The development of the Maine coast has had a crowd-

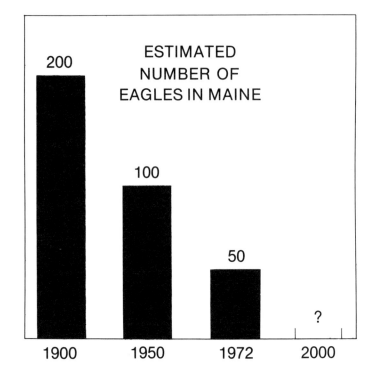

ing effect on the eagle. To a certain extent, the eagle is able to adapt to the presence of humans, but large amounts of stress during the breeding season tend to lower the reproductive rate. A perfect example of this occurred in spring 1972 when a pair of eagles on a Washington County island were harassed and shot at by young boys. The eagles stopped brooding and abandoned the nest.

Problems also arise when a change is made in the habitat around the nest. In an attempt to protect this habitat, the U.S. Fish and Wildlife Service has drawn up an agreement for landowners with eagle nests located on their property. This agreement provides for a protection area around the eagles' nest, limiting excessive human disturbance, including building or use of roads, lumbering, development, etc.

Along with the development of the Maine coast has come pollution of our waterways. These waterways are channels for toxic chemicals; fish populations have decreased, accelerated eutrophication has spoiled lake habitats, and the list goes on. Fortunately for the eagle, the rivers and lakes are being cleaned up, DDT and other toxic chemicals are being banned, fish are being restocked into once polluted waterways, and controls are being put on development, *but* much more needs to be done for the eagle.

Precious wilderness areas necessary for the eagle's existence must be protected and preserved. The laws protecting eagles need to be strengthened and enforced. Most of all, people must begin to realize that if we don't do something soon, there won't be any bald eagles soaring across Maine's skies. They will be seen just as metal ornaments over garage doors or as figures engraved on the President's seal. ∎

Immature bald eagle.

What's the Latest on Maine's Eagles?

By Francis J. Gramlich

MAINE IS FORTUNATE in still having bald eagles in most of their historic territories. In the other New England states and much of the Northeast, nesting eagles are only a memory. Our resident bald eagle *(Haliaeetus leucocephalus)* is truly a magnificent bird. In adult plumage, its white head and tail and contrasting brownish-black of its body are unmistak-able. Adult dress is not complete-ly acquired until the fifth year. Young birds vary from almost completely dark brown in their first season through mottled stages as their mature color gradually is developed. Eagles may weigh 8-12 pounds and their wings span 6-7 feet.

Several studies have shown that shooting has been the major cause of eagle mortality. Today, there can be no excuse for shooting an eagle. All hawks and owls are protected, and eagles especially so. In fact, generous rewards are available for persons furnishing information leading to the arrest and conviction of an eagle killer. At present, an individual could be entitled to $3,500 or more for assistance in bringing an eagle slayer to justice. Penalties of up to $5,000 in fines and imprisonment for one year may be assessed for destroying our national bird.

MANY MAINE EAGLES probably reside here throughout the year although a sizeable number may migrate varying distances in late fall and winter. Availability of food is apparently the controlling factor in movement. In January 1975, a study indicated that perhaps half the hundred-odd estimated population of Maine bald eagles was present along the ice-free portions of our rivers, lakes, estuaries, and coastal areas.

Since eagles are primarily fish eaters, their territories are established along the coast, including offshore islands, inland lakes, and major rivers. Aquatic birds—waterfowl, gulls, cormorants, and others, are also taken when the opportunity is presented. Injured waterfowl are particularly susceptible and frequently become food items in fall and winter. The bald eagle is a relatively poor hunter, and many mammals they eat are in the form of carrion.

Eagles build nests that they use for many years, possibly by successive generations of eagles. Nests are almost invariably located within several hundred yards

Two eagle eggs from Minnesota, placed in foster nests in Maine last year, resulted in one fledged eaglet, pictured below. The author, left, and federal biologist Paul Nickerson hold dead eggs, taken from the same nests, which were later found to be high in pesticide residues.

49

of a large body of water. While eagles may appear at their nesting sites occasionally at any time of year, the birds begin courtship and nesting activities in late February or early March. A territory may contain one or more nests although only one will be actively used to rear young in a given season. A single pair of adults may occupy different nests within their territory in successive years. The nest may be seven feet deep and six to eight feet wide, constructed of sticks up to four feet long and two inches in diameter. A liner of softer materials such as grass or pine needles is included. Eagles most often select white pine as nest trees, but they frequently use other species such as oak, beech, and maple. Without exception, and significantly, the nest trees are in old-growth stands. A factor in the selection of a nest site is the pres-

ence of one or more perching trees in the immediate vicinity. The protection of old-growth stands close to water soon becomes evident in eagle management.

Tolerance of eagles to normal human activity varies considerably. Some will nest fairly close to residences, roads, or agricultural lands, but others seek remote, undisturbed nesting sites. A new road or camp can cause desertion of an established nest. Eagles are most susceptible to disturbance at nest building or at egg laying time, while the same activity at late incubation or after hatching might not bother so much.

They produce their white eggs, about the size as those of geese, in mid-March in coastal areas, somewhat later inland. They take about 35 days to hatch. A normal clutch is two eggs with three be-

ing commonly observed. But, particularly in our areas with declining populations, single-egg clutches occur all too frequently. Seldom is more than one young per nest hatched.

Both parents share in the incubation of eggs and brooding and protection of the young, but the female spends more time at the nest than the male and may do almost all the feeding of the eaglets.

Young eagles are unable to feed themselves until they are half grown. The female tears the prey into small strips and presents the food with her beak. When more than one eaglet is hatched, competition for food is continuous; and if food becomes scarce, the larger, stronger chick gets the lion's share, and the smaller may starve. Normally, by the end of twelve weeks the young eagles are ready to leave the nest. Immature

Before releasing an eagle that had been injured and nursed back to health last summer, Maine and federal wildlife biologists marked the young bird with a colored tag for late identification. The tag can be seen on the shoulder area of the left wing as the bird flies away.

Photo by Jack Swedberg, Mass. Fish and Wildlife

and adults continue to use the nest as a perch and feeding platform after the young birds become proficient in flying.

IT HAD BECOME evident in the early 1960's that eagles were not as numerous in certain areas of Maine as they had been previously. Population and production surveys were begun at that time, and although the intensity of the survey has increased since then, the number of known active nests has averaged close to 35 annually. An active nest is defined as one in which an adult eagle is observed on the nest in an incubating posture.

Maine surveys have been conducted chiefly from aircraft, as eagles have little fear of a plane and will allow a close approach without alarm.

Certainly there are active eagle nests not known to the survey, but suspected locations indicated by the presence of adult eagles during nesting season suggest that there are now probably somewhat fewer than 50 pairs of eagles in Maine during the breeding season.

Production is measured by a later flight, in June, when the young can be easily counted before they are fledged. The production of eagles from the known nests has been disappointingly low—about 0.3 eagles per nesting pair. It has been estimated elsewhere that the minimum production to maintain a stable population lies between 0.5 and 0.7 birds per active nest.

Reasons for the reduction in number of Maine's eagles are not completely understood. Certainly there are more human activities in the eagles' territories. And eagles have been caught in bobcat traps and even muskrat traps. They have even been victims of traffic accidents, and, of course, illegal shooting.

The large-scale use of persistent pesticides and the increase in industrial pollutants have introduced new dangers to our eagles. Analyses of samples from the bodies of Maine eagles and their eggs have shown consistently higher residues of the DDT complex, dieldrin, and PCB's than other eagle nesting areas. In 1974, eggs collected from failed nests showed the highest levels of these materials found since eagle eggs were first analyzed. In 1975, a crushed egg taken from a lower Kennebec nest had a shell 28 per cent thinner than normal. Eggshell thinning has been found to be associated with high levels of pollutants in other birds.

The lower Kennebec nests have been singularly unproductive. No natural reproduction has been known for more than 10 years.

In 1974, with the hope of recruiting young eagles to this population, two eagle eggs were taken from Minnesota and placed in two Merrymeeting Bay nests. Both hatched under foster parents. One eagle fledged and was in the Bay this spring. In 1975, four eggs were transferred from Wisconsin, two of which hatched in their Maine nests.

Sampling of known eagle foods has not yet revealed the source of the residue problem. The pollutants are present, but generally not in excess of residues found in other parts of the country where eagles are still reproducing at a maintenance rate. It is hoped that recent bans on use of persistent pesticides will be reflected in lower residues in our eagles.

THE INDIVIDUAL Maine citizen is becoming more concerned about natural resources including eagles, and intentional destruction of our national bird should become a rarity. Protection of old-growth stands for nesting is an absolute necessity. Our growing system of parks, refuges, and co-operative nest protection areas will assist in maintaining suitable nesting sites. The future is not too dim for Maine eagles, and with a little luck, they may still be flying at our next centennial celebration. ∎

OSPREY

By Kenneth H. Anderson

THE OSPREY, or fish hawk, as it is commonly known, is the largest member of the hawk family in Maine (except for the bald eagle). The male bird is slightly smaller than the female and attains an overall length of 21 to 24.5 inches with a wing span of 54 to 72 inches. The sexes are similar, with the upper parts dark grayish brown, the head and nape marked with white, and the under-parts white. The breast of the female is always spotted or streaked with grayish brown. Due to its large size, the osprey is often mistaken for an eagle. However, it is easily distinguished from the eagle by its white or mottled under-parts. The under-parts of the eagle are always dark.

There is only one species of osprey, and it ranges over the greater part of the world. The subspecies found in Maine — *Pandion hallaetus carolinensis* (Gmelin) — occurs over the entire North American continent with the exception of a narrow zone above the latitude of middle Hudson Bay. Breeding occurs from Florida to Hudson's Bay and Alaska. The winter range extends from South Carolina to northern South America. Osprey have also been recorded in Peru and Paraguay.

ISMAQUES, as the osprey was known to the Indians, is well adapted to its principal occupation of fishing. The body feathers are tough and oily, and the head feathers are extremely heavy to cushion the shock of plunging into the water. The osprey locates its food by soaring thirty to one hundred feet above the water. When the bird spots a fish, it hovers for a moment, folds its wings tightly, and plunges head first into the water, emerging a few seconds later with the fish in its talons.

The Indians felt that this bird possessed the power of luring fish to the surface by means of an oily substance contained in its body. They also believed that if bait was touched with this substance, it would be impossible for the fish to resist it. While the magical powers of the osprey may be open to debate, its fishing ability is unquestioned. The osprey is gifted with a reversible outer talon which allows it to grasp its prey with four opposing talons. The initial grasp is so strong that the bird has been known to have been drowned by its intended victim. Large fish, with the osprey still firmly attached, have been washed up on shore. A Maine fishery biologist witnessed a near drowning of an osprey. The bird, firmly attached to a large fish, was given a merry ride above and below

the water's surface before it could release its talons. The bird floundered in the water, climbed upon a rock, shook the water from its feathers, and flew away none the worse for its experience. However, once the osprey is airborne with its prey, the grip is relaxed, and the fish is turned so that its head is facing the direction of the flight. Usually, the osprey heads for a favorite feeding site to consume its catch.

No preference is shown for any particular species of fish. Carp, suckers, pike, pickerel, perch, tomcod, alewives, menhaden, sunfish, hornpout, and to a lesser extent trout and salmon have been recorded as being eaten by the osprey. Occasionally, a frog or water snake may be taken, but this is the exception rather than the rule.

THE FAVORITE nesting site is usually a tall, dead tree near a large body of water. However, nests are common in salt marshes, on rocky ledges, cliffs, sand dunes, cross-arms of telephone poles, and on abandoned houses. New nests are generally small; they increase in size as new material is added year after year. Knight, in his book *The Birds of Maine*, reports the following measurements of an osprey nest found on an island in Penobscot Bay: outside diameter, 5 feet 10 inches; circumference at the base, 23 feet; height, 3 feet; and depth of inside hollow, 3 inches. It has been reported that a nest on a ledge at the entrance to Pulpit Harbor was used for almost a century.

Two to four eggs which vary in size and in color from white or dusky white to a light reddish-brown, blotched and spotted with varying shades of brown, are laid about the middle of May in the nest constructed of sticks, brush, and rubbish. The young are hatched in twenty-seven to thirty-five days.

UNLIKE the other birds of prey, the osprey is often a colony nester. Palmer, in his book *Maine Birds*, states that in 1866-67 there were ten to twelve nests at a point in the Damariscotta River, and in 1885, thirteen pairs nested at Great Pond marsh in Phippsburg. It has also been reported that at one time there were approximately two hundred nests on Gardiner's Island, Long Island, New York.

Once numerous, ospreys, like many other unprotected birds, were sharply reduced in numbers in the late 1800's. By 1900, the ospreys were rare in the State of Maine. Norton, in a manuscript note published in Palmer's *Maine Birds*, stated in part, "These fine birds paid the penalty of being abundant and solicitous of the welfare of their nest; the birds were attractive targets for the blood-thirsty beings who deemed themselves 'sportsmen' in destroying hawks. Another cause of destruction was the market value of richly-colored bird eggs among collectors; the handsome eggs of this hawk were easy to obtain and brought a few cents in the egg market." However, protection and the old superstition that to kill an osprey brings bad luck have helped to restore this bird to its rightful place in Maine. ∎

Hawks:

FRIEND OR FOE?

The brightly-colored sparrow hawk is commonly
seen hovering over fields where it feeds on insects
and mice. It is Maine's smallest and most abundant hawk.

Unfortunately, very few detailed studies of hawks have been conducted in Maine. Ralph Palmer described the general habits of hawks in his book *Maine Birds.* Howard Mendall wrote a detailed paper on the foods of hawks and owls of Maine in 1944.

Although occasional mentions of the consumption of ruffed grouse, snowshoe hare, and woodcock occur in the various published food studies of hawks, the bulk of the diet includes such non-game foods as snakes, insects, mice, moles, red squirrels, and song birds. Oftentimes, game animals taken by hawks represent unwary and diseased individuals. Frequently, they are animals that are weakened and hence are vulnerable to predation. Therefore, the sportsmen must conclude that the feeding activities of most hawks does not affect the availability of game for the hunter. Nevertheless, hawks are shot frequently and

By Sanford D. Schemnitz

O NE OF THE MOST PERSECUTED groups of birds in Maine is the hawks. Frequently they are considered as vermin and shot (illegally) on sight. Is this harsh treatment justified? Before we can intelligently answer this question, we should consider the hawks' feeding habits and other aspects of their life history.

Hawks are variable in their size, habits, habitat, and behavior. Thus, it is dangerous to generalize. Certain individuals using learned behavior may feed largely on game or poultry while others of the same species feed on non-game animals.

indiscriminately by misinformed hunters under the assumption that local game populations are being benefitted.

In the past, many people classified all hawks as "chicken hawks"—thereby justifying their slaughter of these birds. Admittedly, some species, including the red-tailed hawk, have been known to kill and eat poultry. However, recent changes in poultry husbandry methods have led to indoor housing facilities for chickens in large, multi-floored "chicken ranch" buildings. Thus, hawks nowadays have little opportunity to feed on poultry in Maine.

Hawk populations have shown a precipitous decline in recent years. Undisturbed nest sites are essential for nesting success. In some instances, old-growth, large trees essential for nesting have been cut and harvested.

Another documented cause of hawk declines has been the increased intake of persistent pesticides such as DDT and other chlorinated hydrocarbons used for insect control. Analysis of body tissues from hawks, by personnel of the U. S. Fish and Wildlife Service, has shown increasing amounts of pesticide residues. Also, recent research findings have shown a relationship between these pesticides and calcium metabolism, resulting in various female hawks laying thin-shelled eggs. This causes eggs to crack, and infertility results. The peregrine falcon (duck hawk), a spectacular and skillful flyer which attains a diving flight speed of 200 miles per hour, has shown a recent drastic drop in numbers which seems to be related to cracking of thin-shelled eggs in nests.

HAWKS ARE UNIQUE in many ways. They are renowned for their keen eyesight and are able to detect the presence of a mouse, snake, or other prey by the slightest movement. Unlike most birds, which have side-directed sight, hawks have forward-directed vision. Also unlike most birds, the female hawk is larger than the male.

The shape and size of hawk wings and tail are associated with their flight patterns. Certain hawks—notably the buteos with broad, round wings and fan-

The broad-winged hawk is usually found in wooded areas. Its diet is mainly insects and insect larvae, mice and other rodents, frogs, snakes, etc.

Photo © by Leonard Lee Rue III

shaped tails—tend to soar. In contrast, the falcons—with their long-pointed wings and long tails—are adapted for rapid, flapping flight and high speed dives. Wing and body shape are aids in field identification.

Sharp, pointed talons of hawks are aids in capturing food. The deeply hooked bill is adapted to tearing apart their food prior to swallowing.

Certain admirers of hawks, since the days of the early Egyptians, have captured these birds alive and trained them by reward to pursue and capture game. This sport of falconry is increasing in popularity. (It is unlawful in Maine, however.) Today the hawks with their magnificent power of flight are appropriately the official mascot of the United States Air Force Academy at Colorado Springs, Colorado.

Hawks are variable in their nesting habits. Most species build their nests in tall trees. Adults defend their territory by loud and shrill calls. One species, the marsh hawk, nests on the ground. The sparrow hawk, our smallest and most abundant hawk, nests in cavities in hollow trees or in a man-made substitute, a nest box.

Most Maine hawks are migratory and leave the state in the fall for milder climes. The goshawk and bald eagle are exceptions to the migratory routine and are fairly common winter residents. Other species that are sporadic in Maine in the winter are the red-tail, Cooper's, sharp-shinned, and rough-leg hawk.

Some hawks exhibit distinctive changes in plumage with age. The goshawk, Cooper's hawk, and sharp-shinned hawk are drab brown as yearlings but change to bluish-gray as second year adults. In one species, sex is readily distinguishable. Gray marsh hawks are the males, while brownish marsh hawks are females.

WHAT IS THE LEGAL status of hawks in Maine? They are protected by Section 2466 of Chapter 319 of Maine Public Law. However, it is lawful for the owner or occupant of land to kill hawks *in the act* of destroying poultry. Maine is not unique in legally protecting hawks since 47 of the 50 states now extend legal protection to these birds. But laws alone will not safeguard our hawks. Only a full appreciation and knowledge of the need for their protection by the general public will help to make existing laws successful.

Fortunately, increasing numbers of people are interested in bird watching. These people are avidly concerned with the welfare of birds and especially about the survival of hawks. Nevertheless, if present trends continue, we can conclude that the hawks of Maine face a dim future unless each of us exerts more effort to protect these valuable birds. Also, we need to support programs to regulate the use of persistent chemical insecticides, not only for the sake of hawks but for the benefit of many kinds of birds and mammals including man. ∎

OWLS are a group of birds that have long interested and fascinated man. In North America, there are 18 species of native owls, 11 of which are found in Maine although only 3 species — *barred, great-horned,* and *saw-whet* — occur commonly.

Specific and detailed information on owl abundance in Maine is largely lacking. One measure of the general numbers of Maine owls is the Christmas bird count sponsored by the National Audubon Society. A review of the published Christmas counts for Maine from 1957 to 1972 showed that great-horned, barred, and saw-whet owls totalled 84 per cent of all owls seen by field observers.

In addition to the three common species, there are four winter visitors from the Arctic north country that sporadically invade Maine. Of these, the *snowy* owl is most often seen. Its large size and pale plumage make the snowy owl quite distinctive. The sighting of a *great grey* or a *hawk* owl often creates a great stir of excitement among bird watchers. In the winter of 1973-74, a dark gray, long-tailed, hawk owl perched tamely in trees for several weeks on a farm in Brewer and attracted large crowds of bird watchers for observation and photography. The fourth winter visitor is the *Richardson's* owl, which resembles a saw-whet in size but has a yellow bill and a spotted head.

Short-eared and *long-eared* owls are uncommon residents of Maine. The short-eared owl is a ground nester that prefers open fields and wetlands. Conspicuous black patches on the under side of the wing of the short-ear distinguish it in flight. The long-eared owl is often found in conifers. It is similar in coloration and appearance but about half the size of a great-horned owl.

Two owl species found commonly in warmer climates to the south — the *barn* owl and *screech* owl — are best classified as rare in Maine. The most recent barn owl specimen in the University of Maine scientific collection was found as a roadkill on the Maine Turnpike south of Portland in April 1954. Two nesting barn owls were captured, banded, and released at Westbrook in 1960. The barn owl, often referred to as the "monkey-faced" owl, is also characterized in having long, sparsely-feathered legs. The barn owl is especially prolific, having as many as 11 eggs in a nest. Screech owls, the smallest of the "eared" owls, are unique in that they are the only owl with two distinct color phases — gray and reddish-brown.

DESPITE THEIR VARIABILITY in abundance, owls display certain common but distinctive features and characteristics. Unlike most birds, the female is larger

MAINE'S OWLS

HOOTS WHO?

By Sanford D. Schemnitz

The great horned owl is easily identified by its large size and prominent ear tufts or "horns."

By Sanford D. Schemnitz

*a close look
at a
group of birds
that are
more often
heard than seen*

Its rounded head, brown eyes,
and distinctive pattern of streaks,
bars, and spots identify the barred owl.

in size and weight than her spouse. Egg incubation in most birds begins after the last egg in the clutch is laid, but owls begin incubation after the female lays the first egg. Therefore, the hatching of young owls is staggered and not synchronized, which results in various size fledglings in a nest.

Owls are freeloaders when it comes to nesting, often seeking hawk, crow, or squirrel nests to deposit their eggs. The eggs are plain white, and the normal clutch numbers two to six eggs. The male owl is a model father, assisting in the feeding of the owlets. The fledglings are rather slow to develop and have a long dependence on parental care — up to five weeks in some species. Both parents are often hostile and belligerent in defense of their nest against intruders. Numerous authentic records exist of humans being "dive-bombed" and attacked by owls while climbing trees to examine nests. When disturbed by human

intruders, owls will click their bills repeatedly in a distinctive manner.

Owl longevity in the wild is largely a mystery, but an authentic record exists of a great-horned owl living in captivity for 68 years.

Some owls are exceedingly tame and can be readily approached within a few feet for observation and photographic purposes. The hawk owl, screech owl, great gray owl, and saw-whet owl show little fear of man, and saw-whets can often be captured by hand.

The flight of owls is silent, and most of them hunt at night. Those that feed in the daytime include the snowy, short-eared, and hawk owls. Owls exhibit a varied diet, most often feeding on whatever prey is most readily available. Research studies show that many owl populations fluctuate in relation to rodent abundance. When food is lacking, some Arctic owl species do not nest, or else their offspring succumb to

The saw-whet owl is the smallest of Maine's native owls. This one chose one of the Department's duck nesting boxes as a good place to raise her own young.

starvation. Interspecies competition between owls results in large owls feeding on small ones. Owls seldom kill prey in excess of their immediate food needs. They swallow their prey in large chunks, digest the flesh, and then regurgitate the cleaned bones and fur or feathers in a pellet form. Owl presence can often be detected by an accumulation of these pellets and fecal splashing on the ground beneath roost trees, which are often used repeatedly.

OWLS AND HAWKS are collectively called *raptors*. Most hawks in Maine are migratory in contrast to owls, many of which are resident. Owls resemble hawks in having hooked bills and powerful feet with long, sharp talons specialized for grasping, killing, and carrying prey. They are distinguished from hawks by their soft, fluffy plumage, seemingly neckless heads, and forward facing eyes.

Although, like humans, owls have their eyes in the front, their eyes are in a fixed position so they have to move their heads for side vision. This binocular vision gives the birds a three dimensional sight ability necessary for distance determination essential for hunting live prey. Unlike hawks, most owls are nocturnal. Owls' eyes contain more rods than cones, adapting them to acute night vision. Another feature that all owls have in common is their acute sense of hearing which is used as an aid in locating prey.

Owls are difficult to observe because of their nocturnal activity. And in the daytime, they often retire to dense forest vegetation where they remain inactive. Fortunately for the bird watchers, the shy and retiring owls tend to be vocal, particularly during the breeding season, which begins in March for the great-horned owl in Maine. Calling is part of the courtship ritual and facilitates the defense of the nesting territory, thereby restricting intruding owls. Oftentimes a reasonably good, human imitation of an owl call at dusk or daybreak will elicit an answer from an owl deep in the dense forest.

The call of Maine's three common owl species is distinctive and readily recognized. The barred owl calls frequently and seems to say "who cooks for you, who cooks for you-all?" The great-horned owl's hoot or call consists of a series of consecutive "hoo hoo hoooo." The single note of the tiny saw-whet owl resembles a saw being sharpened with a file.

LET US CONSIDER in a little more detail the three common owls of Maine. The great-horned owl is often referred to as the "tiger of the air." A good-sized female will have a wingspread of four feet and weigh four pounds while her mate averages three pounds.

One of the frequent items in the diet of the great-horned owl is the striped skunk. In Maine, this bird is often attracted to garbage dumps where it feeds on rats. This owl species is the first to nest in late winter.

The barred owl is a general feeder — consuming a variety of birds, mammals, snakes, and insects. Unlike the other Maine owls with light-colored eyes, the barred owl has dark brown eyes. Barred owls are often found near lake shores or rivers.

The smallest of our native owls is the saw-whet, weighing four to five ounces. The saw-whet is the only owl with a distinct immature plumage that differs from the adult feather coloration. First year saw-whets are a solid color, not streaked and normally chocolate brown, and they have conspicuous white "eyebrows." The adult birds are light brown and streaked in appearance. Saw-whets prefer conifer forests and often nest and roost in old woodpecker holes. Bird watchers often locate this owl by tapping with a stick on hollow trees with nest holes.

A QUESTION OFTEN ASKED about the diet of owls is "Are the birds beneficial or harmful?" Again, the answer is not simple since the group is most varied in feeding habits. Numerous detailed food studies show that non-game birds and mammals predominate in the diet although this is not to say that a large owl will not take a grouse, pheasant, rabbit, squirrel, or other game animal. Interestingly, individuals within a population will develop a taste for particular prey, some-

times feeding exclusively on a certain species of mammal or type of bird. Others more typically feed on a broad array of different prey.

Owls have few enemies other than man. They are often harassed and pestered by crows, and as mentioned, some owls prey on other owls. Deaths also occur from collisions with utility wires and motor vehicles along highways.

What can we do to help increase owl populations? First and foremost, we need to publicize that all owls are completely protected by state and federal laws. Support of protective laws by the general public is essential to perpetuate owls. One management practice to aid owls involves the building and placement of nesting boxes in woodland areas. Care must be taken to roughen the inside or to provide a wire ladder so that the young owls can climb up to the hole when they are ready to leave the box. A horizontal sun perch near the entrance hole also enhances box usage. Owls are a valuable and interesting avian group that we should all strive to encourage. ■

The snowy owl is a sporadic winter visitor to Maine. This large, white owl prefers open country and is a daytime feeder, unlike most other owls.

© Leonard Lee Rue III

Common loon on nest.

The Lonely Yodeler of Maine Lakes

By Tom Shoener

CHANCES ARE GOOD that anyone who has ever fished or camped at a Maine lake can recognize a loon, at least by sound, if not by sight. These large, water birds are so distinctive that they would be im-

possible to ignore, mistake, or forget. But aside from their striking black and white coloration, their prolonged underwater swimming ability, their curiosity, and their wild cry, how much do you really know about loons?

Did you know that in Maine we have two species of loons? The red-throated loon is mainly a coastal bird and is rarely seen except on spring and fall mi-

gration. The common loon, sometimes called "the great northern diver," is the one we see on lakes and larger ponds from early spring through fall. Although the common loon is larger than the red-throated loon and is colored differently, much of what may be said about one species is also true of the other.

Loons are long-bodied birds with short tails, thick necks, and strong, dagger-like bills. They swim lower in the water than ducks and geese, more resembling grebes, although grebes are smaller and have thinner necks.

Loons are somewhat primitive birds that show no close affinities to any other bird orders. They are among the very few birds whose bones are solid and heavy. Their specific gravity is close to that of water, and they have the ability to expel enough air from their bodies and from under their feathers to be able to sink slowly and quietly beneath the water surface.

The common loon is a large bird, about the size of a goose. An adult in breeding plumage in the spring bears white marks arranged in regular lines across its black back and wings. Its head and neck are a glossy purple-black. The bird sports a white "necklace," and its upper throat bears a thin line of white spots. The loon's eyes are red, and its three-inch bill is black. The breast and underparts are white. Both the male and female are similarly marked.

Adult common loons lose their breeding plumage in late summer, becoming dull gray-brown with a dingy white throat. Even their eyes change color, from red to brown.

Loons' legs are placed far back on their bodies, and they are the only birds whose legs are encased in the body down to the ankle joint. As a result of this peculiar leg arrangement, loons can scarcely walk on land. The best they can do is waddle awkwardly in an almost upright position or push themselves along on their breasts and flopping wings.

Loons have one of the smallest wing areas of all birds in proportion to the weight of their bodies. With such small wings, they cannot fly from land at all, and they require a long, splashing run across the water before they can finally rise into the air. Once airborne, however, their flight is strong, swift, straight, and long sustained. In flight, loons appear hump-backed

and pointed at each end. The head and neck are carried below the level of the back, and the large feet trail slightly downward.

Almost as much water runway is required for a loon landing as for takeoff. Somewhat less than graceful, the landing can perhaps best be described as a controlled crash. The loon, unable to check its momentum with its small wings, flies closer and closer to the water until it finally splashes in, breast first, plowing a long, watery furrow.

What loons lack as walkers and fliers they more than make up for by their abilities in the water. They are rapid swimmers, using powerful thrusts of their large, webbed feet. As divers, they are probably unequalled. The birds plunge forward with arched necks to dive quickly and gracefully beneath the surface, creating hardly a ripple.

They can remain underwater for lengthy periods, dive to great depths, and travel long distances submerged. Normal feeding dives are less than one minute long, but escape dives of two to three minutes are not uncommon. Some dives of eight to 10 and even 15 minutes have been reported, but these times are almost impossible to authenticate because of the loon's ability to take a brief breathing spell unobserved, by swimming with only its bill protruding above the water surface. Loons have been caught in fish nets 60 feet beneath the surface, and dives to depths of more than 200 feet have been reported.

Loons feed largely on fish, supplemented occasionally with such animal life as frogs, leeches, insects, and shellfish. Fishermen are sometimes apt to blame loons for a luckless day on a trout pond. Although it is a fact that they dine on desirable game fish as well as other species, it is unlikely that the fish population in any pond large enough to accommodate a pair of loons could be seriously depleted by them. It's also unfair to blame loons for competing with us for a

meal of trout. They have a right to live, too, and they're an important part of the total atmosphere of a wilderness pond.

LOONS SPEND THE WINTER mainly on salt water, south as far as Florida and the Gulf Coast. They are not uncommon along the Maine coast in the wintertime, but because their winter plumage is so different from their summer coloration, they might not be recognized as loons.

The breeding range of the loon is in Canada and the northern states. In the spring, they migrate through Maine just behind the melting ice. Those that remain to nest here select lakes and larger ponds throughout the state except close to the coast and in extreme southwestern Maine where they are less common as nesters. Nesting loons prefer solitude, and some former nesting lakes have been abandoned as too many motorboats, cottages, and people came on the scene.

Not only do loons prefer absence of human intruders on their nesting territory, they also don't like other loons there. Ponds and small lakes never hold more than one pair of breeding birds. Larger lakes may have several pairs but always at some distance from one another.

Loons are paired when they arrive at their chosen nesting lake, and it is believed that they mate for life. Nesting normally starts in late May and early June in Maine. The nest itself is usually made of aquatic vegetation and is often located on the end of a point that juts into a lake or on an island, a muskrat house, or on a floating mat of vegetation. Nests average about three inches deep, with an inside diameter of a little more than one foot and an outside diameter of about two feet. They are always built within a few feet of the water so that the birds can slip out of the nest and directly into the water. The same nest sites are used year after year.

Loon chicks leave the nest within minutes of hatching and can swim immediately.

Loons typically lay one clutch of two eggs each breeding season. Occasionally, only one egg is laid, and rarely there are three. They are about three and one-half inches long. Coloration varies from dark olive-green to brown, with a sprinkling of small, dark spots.

Both prospective parent loons take turns sitting on the eggs through the 29 to 30 day incubation period. They are rather easily frightened from the nest early in incubation, but as hatching time nears, they become more reluctant to leave. The parents almost never leave their eggs alone voluntarily.

Loon chicks are covered with thick, brown-black down. They leave the nest within minutes of hatching and can swim at once. During their first few weeks, the young frequently ride on the parents' backs among the feathers.

For a month to six weeks, the adult loons and their offspring stay in the vicinity of the nest, gradually increasing their cruising radius as the chicks' swimming and diving abilities improve.

The baby loons eat regurgitated food, minnows, crustaceans, and bits of vegetation, supplied by the parents. Some observers report that although both parents gather the food, only one does the actual feeding of the young. The chicks learn to dive and catch fish while still quite young, but it is usually about six weeks before they are able to catch most of their own food.

Young loons grow rapidly, lose their down, and acquire gray feathers. By late September, they can fly and are almost as large as their parents. The family groups usually begin breaking up once the young have learned to fly. Fall migration occurs in October and November, with the young birds usually preceding the adults by several weeks.

The immature birds do not migrate back to nesting areas in the spring. Maturity and adult coloration are not attained until their third year, and until then the immature birds congregate in the summer on coastal waters.

LOONS ARE PROBABLY KNOWN BEST by their calls, which are loud, varied, and — for the uninitiated camper spending his first night on the shore of a wilderness lake — rather frightening. They seem to have an uncanny ability to do most of their calling at times when its effect is most dramatic — in the stillness of the night and predawn and during the period when all else in nature is quiet before an approaching storm.

Loon calls have been variously described as wild, crazy, eerie, uninhibited, cacophonous, weird, maniacal, screaming, laughing, wolf-like, and mournful. Regardless of how you describe the sound, you'll probably agree with the author who long ago asked, "Who has ever paddled a canoe, or cast a fly, or pitched a tent in the north woods and has not stopped to listen to this wail of the wilderness? And what would the wilderness be without it?"

Although there are a great number and variety of loon calls, four basic types of calls have been described: (1) the "tremolo," (2) the "yodel," (3) the "wail," and (4) the "talking calls." The following information on loon calls is condensed from a study entitled *The Common Loon in Minnesota* by Sigurd T. Olson and Dr. William H. Marshall, published by the University of Minnesota.

The tremolo is the best known loon call. Fundamentally, it consists of from three to eight or ten notes uttered rapidly, either as a medium or high-pitched tremolo, depending upon the individual loon and purpose of the call. The tremolo call is given under varying circumstances and may be regarded as an all-purpose call, registering alarm, annoyance, worry, greeting, and courtship.

The yodel call in many respects is similar to the yodel effected by the human voice. The call and its infinite variations have a definite throaty quality. This is the weirdest and wildest call of all and has been described as beautiful and thrilling, and also as maniacal and blood-curdling.

The yodel is most typically heard at dusk, during the night, and in the early morning, as an integral part of choruses. Often it is given alone and seems to stimulate the tremolo call, later followed by similar yodeling. Loons flying at dusk and at night are prone to break out with a wild series of yodels, swinging in wide circles, calling repeatedly, inciting other loons on the lakes to further calling. The final result is a pandemonium of sound, coming from miles around. This may die gradually, or it may stop abruptly as though by a signal, which makes it all the more impressive.

The third basic call, the wail, is often mistaken by the inexperienced for the howl of a wolf. This call is given with the bill almost closed, the throat swelling considerably as the sound is seemingly forced out of a nearly closed bill. It often prefixes the yodel as well as the tremolo. It is often used as a summons to the mate or young.

The talking calls are simple, often one-syllabled notes, used as communicative utterances between mates or members of a flock. They are never given in excitement or as part of courtship display. The "hoot" is the most common of these calls. It consists of short, abrupt, medium-pitched, one-syllabled notes uttered at irregular intervals of from three to ten seconds. A second talking call is the "kuk" call uttered during flight. Other,

barely audible, sounds are exchanged between a pair of loons.

ONE DOES NOT observe loons very long before becoming fascinated with their behavior patterns. Some of their more dramatic displays are associated with courtship, protection of young, and defense of territory.

Courtship displays are often characterized by such activities as bill dipping, head flips, quick dives, exaggerated preening and stretching, and long footraces across the water, usually accompanied by repeated high tremolo calls. One such race, after but a moment's rest, is followed by another and then another, over and over again. Footracing is also a favorite activity of parent loons and their young.

Loon parents are very devoted to their offspring, and they employ several tactics for protecting them from intruders. A favorite is the decoy act in which one parent calls loudly and makes itself as conspicuous as possible to the intruder while the other parent tries to sneak from the area without being observed.

It is in defense of their territory that loons put on their most spectacular performances. When territory defense is heightened by protection of eggs or young, these frenzied displays sometimes border on violence.

Daily flocking is another interesting trait of loon behavior. Early in the summer, birds believed to be unsuccessful breeding pairs, non-breeding pairs, and unmated birds begin to associate in flocks which build up through the day, reaching peak size by mid-afternoon — sometimes 20 or more birds on larger lakes. These flocks begin disintegrating late in the afternoon, and by late evening the individuals are completely dispersed. The purpose of flocking is unknown although some observers suspect that it may be nothing more than a gathering at good fishing grounds.

LOONS ARE LONG-LIVED. In a clean and undisturbed environment, individuals have been known to survive for 20 to 25 years. Enter man: Oil on coastal wintering areas...pollution in lakes...insecticides in the food chain...shooting...harassment by boaters...destruction of nesting sites...fluctuating water levels during the nesting season...all are either known or strongly suspect factors that shorten individual loon lives, limit loon populations, or restrict their range.

Despite these problems, loons are certainly not facing extinction. Some of their present difficulties should diminish in importance in the future as new environmental protection laws begin to show their value. The various pollution and pesticide laws, and shoreline and marshland protection efforts, will benefit loons and man alike.

Loons have been protected from shooting for some time by both federal and state laws, and there is really no excuse for "mistaken identity" shooting during the waterfowl hunting season. Boaters can help loons by steering clear of nesting areas during the early summer breeding season.

One of the biggest threats to loons in the future will be the continued nibbling away of the private nesting places that they require. As lakes are developed, and humans in increasing numbers make their presence felt, the breeding range of loons in Maine can be expected inevitably to shrink.

Fortunately, Maine still has hundreds of lakes and ponds where man's impact is not yet too severe, and the wild enchanting beauty of the loon and its call continues to be an inseparable part of the adventure of fishing and camping on lonely lakes. ∎

THE MOURNING DOVE

By Warren A. Eldridge
Game Biologist

AN INCREASINGLY common sight along Maine roadsides and around back-yard feeders is the mourning dove, or *Zenaidura macroura* if you prefer scientific names. Mourning dove sighting reports have come in from as far north as Presque Isle, and in some southern Maine areas the birds have actually become so numerous that they are no longer a curiosity.

The mourning dove belongs to the same family as the common pigeon and resembles this familiar bird in many ways. The main differences are the dove's smaller size, slimmer outline, and long, pointed tail. Unlike pigeons, most doves migrate south in the winter, leaving around September and returning in March or April. A few hardy doves usually brave the Maine winter, and flocks of from 5 to 30 birds are sometimes seen around feeders, primarily along the southern coastal portions of the state.

Both sexes have a gray-brown back and wings. The abdomen in both is a creamy-yellow. Tail feathers are gray, the lateral ones having conspicuous white tips. The adult male has a light blue cap on its head; the smaller female has a grayish cap similar to the color of the back and wings. The male's breast is tinged with pink, while the female's is light tan. Juveniles can be distinguished from adults by their smaller size, mottled coloration, and white-tipped primary coverts overlying the large wing feathers.

The mourning dove nests in all of the contiguous forty-eight states. Preferred nesting sites include stands of evergreens and heavily-foliated deciduous trees adjacent to fields and a water supply.

After choosing a nesting area, the male begins cooing to attract an unmated female. The coo is soft and five-syllabled and is a familiar springtime sound wherever mourning doves are common. The female is capable of cooing but does so infrequently, and the sound cannot be heard at a very great distance.

The mourning dove nest is quite flimsy and is constructed of small twigs brought to the female by her partner. Ground nesting is not common but does occur.

The eggs are white and are laid in clutches of two. They can often be seen through the bottom of the loosely constructed nest. During the two-week incubation period, the male is usually on the nest from morning to late afternoon, at which time the female returns, allowing the male to feed before sundown.

The young doves, or squabs, are naked when hatched, but grow rapidly and leave the nest in about two weeks. They are fed and protected by both parents. For the first few days after hatching, they are fed a combination of regurgitated seeds and "pigeon milk" which is composed of cells sloughed off from a milk gland in the adult's crop. Later, the diet consists mainly of seed material. After they leave the nest, the young are on their own.

In northern states, each pair of doves may raise three or four broods each year; however, this high reproduction rate is offset by high nesting losses and juvenile mortality.

Mourning doves are almost entirely seed eaters; any insects and bits of plants they eat are usually picked up accidently. Being unable to cling to upright stems or scratch for food, doves must rely on seeds that have fallen to the ground. For this reason, they often feed in seeded or harvested fields. In areas of high dove concentrations, they can do damage to grain crops; however, much of their diet is composed of weed seeds, making them quite beneficial.

Just before sundown, doves will flock together and fly to watering sites where they drink enough water to last them overnight. Doves have the ability to suck water up through their bills, rather than having to toss their heads back to swallow each bill-full. This adaptation allows them to take up large quantities of water in a short time. Following the feeding and watering periods, they return to evergreen stands to spend the night.

Although protected in Maine, the mourning dove is the most important single game bird species in North America from the standpoint of annual harvest. About 41.9 million doves were taken in 1965 in the United States alone. Few birds have greater combined sporting and asthetic appeal.

Even though the number of doves appears to be on the increase in Maine, much is yet to be learned about their numbers, productivity, distribution, and migration habits. Any readers with information on mourning doves, particularly nesting birds, are urged to notify the Game Division office at Nutting Hall, University of Maine at Orono, or the Maine Audubon Society at 57 Baxter Boulevard, Portland. ∎

Crows and Ravens –

Nature's All-purpose Birds

DID YOU EVER have a bird outwit you? If you did, it most likely was one of those "black rascals," as they are called by some: the crows and ravens. Probably no bird is steeped with as much legend and associated with as many "fowl deeds" as these two members of the family Corvidae.

The raven has always been considered something sinister and evil. He was as well known in the Old World as he is here. His jet black and sometimes irridescent feathers and his superior intelligence among birds have helped him win this reputation. Throughout folk tales, he was regarded as mysterious and semi-sacred. His ominous sounding calls were believed prophetic and to

By Peter A. Cross

portend evil. When the Pilgrims landed here, the raven was all too familiar to them, and he quickly became known as a killer of sheep and other animals. Man continued his war with him and gradually reduced his numbers.

The crow, while not having the stigma of evil attached to it like the raven, has always been considered a rascal or a scoundrel by man. The crow tends to live closer to man than the raven, perhaps finding things easier there. He quite often comes into conflict with the farmer while trying out some of his corn or maybe even killing some poultry. He, also, has a good ability to get men angry at him. Crows are almost impossible to sneak up on because they always post a guard while eating and seem to have learned the range of a shotgun and how to stay just outside that range. It has been

said that if men wore feathers, few would be clever enough to be a crow.

Both the crow and the raven are completely black and are large birds. The crow averages about 18 inches long and has about a three-foot wing spread. His call is the familiar, "caw, caw, caw." He is probably among the best known of all birds. The raven looks quite similar but is larger. He averages 24 inches long and has a four-foot wing spread. His bill is coarser and stouter than the crow's. His voice is much different, also, having been described as a hoarse "cronk" or a loud "craak-craak." These differences are important to the hunter as there is a completely open season on the crow, while his brother the raven is on the protected list, and it is illegal to kill one.

The crow and the raven are found over wide ranges. The crow is more an eastern and central North American bird, staying in areas where man is found. In Maine, the crow is widely distributed in the summer except in the deep woods; it winters chiefly along the coast. The raven is a bird of the north and boreal forests. It is found in Maine along the northeast coast and in the interior.

The crow and raven both nest in the spring and lay 4 to 7 eggs. The large nests are often found in the tops of large conifer trees, and the raven often nests on rock ledges and cliffs. After incubating the eggs for 15 to 18 days, the parents have a brood of constantly hungry young to feed in the nest for 4 to 5 weeks. The diets of crows and ravens are pretty much alike. They are completely omnivorous, meaning they will eat anything. They feed on dead animals, live animals, nuts, fruit, corn, eggs, young birds, insects, and mollusks. They can be very helpful in keeping down insect and mouse populations around farms although they occasionally get into trouble eating crops or wild and domestic animals and their young.

Next time you drive along a road and see a big black bird in the road, think of him as a mixed bag of blessings and drawbacks but interesting to have around. And don't worry about hitting him with your car: he's too smart for that to happen!

 52089

B.H. BARTOL LIBRARY